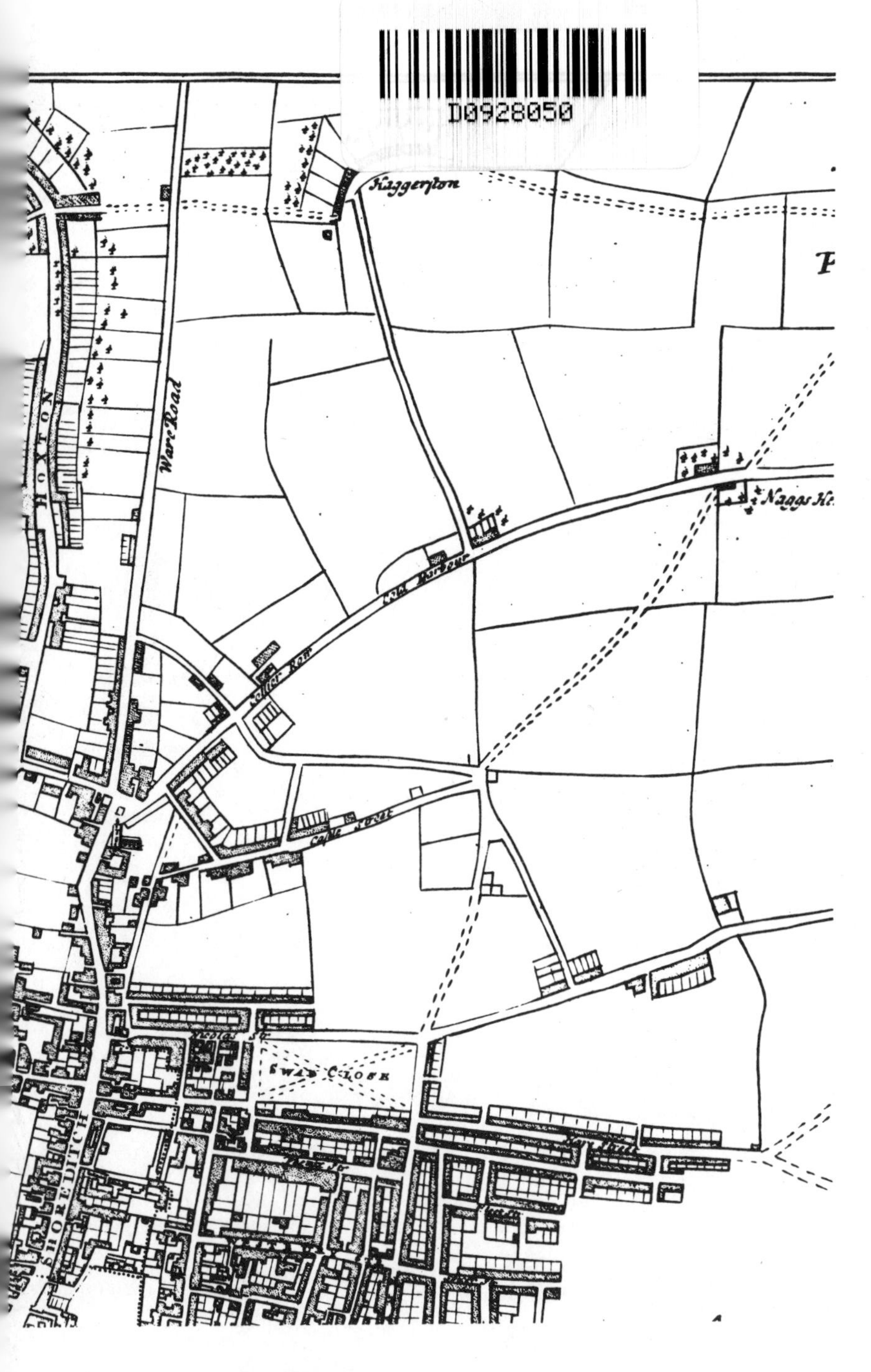

WARK

The Ingenious Mr Fairchild

Also by Michael Leapman:

One Man and his Plot
Barefaced Cheek
Last Days of the Beeb
Treachery?
Kinnock
The Book of London (ed.)
Companion Guide to New York
Yankee Doodles
London's River
Treacherous Estate
Master Race (with Catrine Clay)
Eyewitness Guide to London
Witnesses to War

The Ingenious Mr Fairchild

THE FORGOTTEN FATHER OF THE FLOWER GARDEN

MICHAEL LEAPMAN

HEADLINE

First published in 2000
by HEADLINE BOOK PUBLISHING

Michael Leapman would be happy to hear from readers with their comments on the book at the following e-mail address:
mhleapman@email.msn.com

10 9 8 7 6 5 4 3 2 1

British Library Cataloguing in Publication Data

Leapman, Michael, 1938-
Ingenious Mr. Fairchild : the forgotton
father of the flower garden
1.Gardeners - Biography
2.Beds (Gardens) - History
I.Title
635'.092

ISBN 0 7472 7359 6

Endpapers: Survey of London, Westminster and Southwark (1700)
(Guildhall Library, Corporation of London)

Typeset by Letterpart Limited, Reigate, Surrey
Printed and bound in Great Britain by Clays Ltd, St Ives plc.

HEADLINE BOOK PUBLISHING
A division of the Hodder Headline Group
338 Euston Road
London NW1 3BH
www.headline.co.uk
www.hodderheadline.com

CONTENTS

ACKNOWLEDGEMENTS

Authors customarily pay tribute to their partners and family for putting up with them during the months of preoccupation with the work in hand. My debt to my wife Olga is of a different magnitude. Quite simply, I could not have written this book without her, for she has done most of the gruelling research in libraries and record offices, allowing me to get on with the less demanding business of putting down the words. My gratitude to her is boundless.

Early on, while reading some correspondence about Thomas Fairchild in the office of the Clothworkers' Company, I came across letters from the late Judge Gilbert Leslie, a former liveryman of the Gardeners' Company, who had been researching Fairchild with a view to writing a book about him. Sadly he died before he could complete it, but his widow kept his notes and materials and, in an immensely generous gesture, she and her daughter Caroline allowed me to borrow them while I was writing this book. The material was especially useful on the botanical aspects.

After putting a notice in a local magazine in Aldbourne I was contacted by Julia Hunt, who has family links with Stephen Bacon, Fairchild's nephew and heir. She had done a body of meticulous work on her family history and selflessly agreed to undertake further research on our behalf. We are grateful to her and to Ann Currie, who provided useful material on the history of Aldbourne.

Several academic institutions gave willing help. Niki Pollock of Glasgow University Library not only sent materials but also put us in touch with Dr Helen Brock, who has done a detailed study of the botanist Dr James Douglas and kindly lent us her handlist of the University's collection of his papers. At the Royal Society I consulted the records of Fairchild's two appearances there, as well as the Society's long connection with the Fairchild Lecture. I spent a profitable day at the Plant Sciences Department and the Sherard Herbarium at Oxford University. The National Portrait Gallery and the Witt Library at the Courtauld Institute filled in background on Richard van Bleeck, who painted Fairchild's portrait. The University of Virginia sent copies of some Mark Catesby material.

The efficiency of the new British Library was appreciated by both of us as we consulted original manuscripts and books. Staff at the Royal Horticultural Society's Lindley Library and the London Library were of great assistance. The librarian at Winchester Cathedral kindly allowed me to look at its records relating to seventeenth-century Aldbourne. We examined other material at the Public Records Office at Kew, the National Register of Archives, the Natural History Museum, the Guildhall Library, the Corporation of London Records Office, the London Metropolitan Archives, the Wellcome Institute, the Family Records Centre, the archives department of Hackney Borough Council and the County Record Offices of Wiltshire and Berkshire, where Yvonne Cocking lent an experienced hand.

Officers of the Worshipful Company of Gardeners took great interest in the project and provided valuable input. Rodney Petty, the assistant clerk, and John Schweder, a senior past master, were particularly generous with their time. At the Clothworkers' Company, its archivist D. E. Wickham willingly put all his material on Fairchild at my disposal, and we also had help from the Drapers' Company.

The book arose out of a discussion I had with Heather Holden-Brown, non-fiction publishing director at Headline, at the home of my agent Felicity Bryan. Both Heather and Felicity have been immensely supportive throughout, as has Celia Kent, managing editor at Headline, who has skilfully and sympathetically nursed the book through to publication. Many others have helped in a variety of ways at every stage. My thanks to all.

LIST OF ILLUSTRATIONS

Introduction

As the man who made the first recorded flower hybrid in Europe, some time before 1720, Thomas Fairchild clearly rates more than the footnote he usually receives in books about botany and garden history. After all, most of the flowers that we grow in our gardens are hybrids, crosses between species that, over the years, have been selected and refined to provide an infinite range of colours and shapes, and to perform well in the highly artificial conditions we customarily create for them. Fairchild made no pretence of being a scientist, but he was aware of the significant advances being made in understanding plant sexuality and reproduction, and was the first to put into practice what had until then been no more than theories. He did not discover sex in plants, but he showed the world how to exploit it.

Yet his claims as an important horticultural innovator went largely unrecognised for more than a century, because nurserymen and gardeners were slow to understand the enormous benefits that hybridisation would bring in increasing the range of plants they could offer to their customers and patrons. They preferred the old methods of careful selection of colour and form, and of propagation by cuttings and well-established grafting techniques. Moreover, intrepid plant-hunters had begun to bring in growing numbers of novelties from Asia and the Americas – sources that showed no sign of drying up. The gardening world

Thomas Fairchild's Family

John + Elizabeth Butt of Lambourn

John + Marjorie Fairchild of Aldbourne

Richard Butt b. 1631

Thomas Shepherd d. 1664 = 1659 (1) Ann Butt b. 1636 d. 1679 Aldbourne

3 other children

1665 (2) = John Fairchild (Vairchilde) of Aldbourne d. 1668

1670 (3) = John Bacon of Aldbourne d. 1713 = 1680 Mary

6 children

4 daughters

Tymothie Fairchild d. 1710 = Margaret Plow d. 1730

Richard Butt

3 daughters

Ann Shepherd 1660–1665

John Fairchild b. and d. 1666

Thomas Fairchild b. 1667 Aldbourne d. 1729 London

John Bacon 1671–1741 = 1696 Mary Collins

Stephen Bacon 1674–1753 Aldbourne = 1703 Martha Archer

Ann Bacon b. 1676 = 1700 John Woodley

Timothy Fairchild 1678–1688

John Fairchild b. 1683

3 children

William Hill = [2] Mary Baynham d. 1753 = [1] 1730 Stephen Bacon b. 1708 d. 1734 London

John Bacon b. 1712 d. 1729 London

2 daughters

Thomas Woodley

Ann Woodley

Lydia Woodley

3 children

Fairchild Bacon 1732–1754 London

(1) (2) (3) marriages of Ann Butt

[1] [2] marriages of Mary Baynham

became convinced of the value of Fairchild's technique only when hybridisers began producing a dazzling new array of popular plants and flowers, from rhododendrons to sweet peas.

This initial caution in continuing the work Fairchild started was in part to do with a widespread reluctance to interfere with the natural processes by which plants breed. This is the same inhibition that drives many of today's opponents of the genetic modification of plants.

Real concerns about the safety of these new techniques, especially when they are applied to food crops, have become entangled with the mystical belief that somehow it is wrong for humans to act in a godlike manner and arrange the natural world according to their own blueprint. It quickly became clear to me that Fairchild was troubled with similar doubts.

Very soon after I began researching this book I understood why, until now, no full-length biography of him has been attempted. Original documents relating to his life are scarce. He was a practical nurseryman, not a scholar or a wealthy aristocrat, so he had neither staff to collect his papers nor a library in which to store them. Any bills or other documents connected with his nursery at Hoxton, North London, have disappeared, presumably long before the nursery itself was built over in the mid-nineteenth century. His name crops up in other people's correspondence, but no letters of his own have been traced, apart from one or two that he wrote for publication by his contemporary, Richard Bradley. As the archivist E. J. Willson told a conference of the British Records Association in 1975: 'In the mind of every worker in this field there is the persistent thought that somewhere many more letters to and from nurserymen may have been preserved – but how to discover them?'

Two appearances by Fairchild at meetings of the Royal Society are documented, and at one of them he read a paper that survives in his own hand. A few dried and pressed examples of the flowers he grew – such as the one pictured on the cover – were preserved by contemporary botanists, some of whose collections are still intact. He wrote a charming book called *The City Gardener*, the first manual devoted to the difficulties, along with the pleasures, of making things grow in crowded, smoky London. There is a portrait of him in the Department of Plant Sciences at Oxford University but no indication of the occasion for which it was painted.

Trying to put flesh on this shadowy horticultural pioneer has proved an absorbing if often frustrating experience. As a journalist by trade and inclination, this is the first book I have written about a historical character – in other words, someone who cannot speak for himself, even to say 'no comment', and who has no friends or colleagues to whom I can turn for insights and anecdotes. For the first time I have had to carry out my research not primarily on the telephone or in newspaper cuttings files but in libraries full of ancient documents. Fortunately, my wife Olga is not only adept at finding her way through the resources of such institutions but also enjoys working in them, and she has done the bulk of the detailed research for this book. The temptation to draw elaborate conclusions about Fairchild's life from the limited facts that she and I have managed to unearth is one that I suppose historians must constantly be faced with. I hope we have resisted it.

It soon emerged that, although archive material about the man himself was scarce, there was a wealth of illuminating documentation on the circles in which he moved and the state of scientific research at the time. In Chapter 3, I quote extensively from the correspondence between James Petiver and Richard Bradley, who was in the Netherlands collecting plants for Fairchild but was constantly running out of money. To earn some, he passed himself off as a doctor and persuaded Petiver, an apothecary, to send him some recipes for medicines.

The letters are astonishingly direct and richly entertaining, but they also show the precarious circumstances in which scientific research and experiment had to be carried out when few formal organisations promoted the work and centres of learning were, in the main, rigid in their exclusive devotion to classical studies. The Royal Society was founded to help fill this gap by enabling innovative scientists to exchange ideas and discuss the results of their experiments, but for many of its members science was no more than a hobby and they required independent means, which Bradley did not have. My account of the dissection of a dead elephant on the lawn of Sir Hans Sloane's house in Chelsea is an example of the talent for improvisation that scientists had to deploy if they were to make progress in understanding the workings of the natural world.

From references to Fairchild in these letters and in other documents there emerges a picture of a modest, patient and accommodating man, immensely respected for his knowledge and his meticulous methods, always willing to put himself out for his friends. He did a great deal of work for Bradley, conducting experiments for him and providing lists of plants for publication, and it would be surprising if he received payment for these services, even if he sought it. Bradley was a quarrelsome man who fell out with many botanists and gardeners of his time – especially Patrick Blair – but the imperturbable Fairchild seems to have stayed on good terms with all of them.

He involved himself in the affairs of both the Gardeners' Company and the Society of Gardeners, which he helped establish along with Philip Miller of the Chelsea Physic Garden. He came to the aid of the head gardener at Charterhouse, between Hoxton and the City, who was trying to squeeze more money out of his employers to keep the gardens in shape. He clearly regarded his fellow Hoxton nurserymen not as competitors but friends. The feeling was mutual, for one of them, Benjamin Whitmill, gave his son the Christian name of Fairchild and another, William Darby, appointed 'my very good friend Thomas Fairchild' as his executor. Fairchild in turn named the Hoxton nurseryman Richard Spier as one of his executors. His good relations with plant-hunters resulted in their supplying him with a steady stream of novelties, and in return he would send them plants to introduce overseas, although there is no record of his ever travelling abroad himself. The plant-hunter and illustrator Mark Catesby, who worked in Fairchild's nursery, was a witness to his will.

The will provided clues both to his family circumstances and to his ambivalent attitude to the great botanical advance he had made. In leaving money for a sermon to be preached at his parish church he was not breaking new ground: it was a common form of public piety in the period. But the two alternative subjects he stipulated for the sermons were revealing, stressing as they did the supreme role of an all-knowing God in the creation of species, despite evidence – including his own experiments – that could be taken to suggest otherwise. At a time when religion was a prime cause of warfare and dispute, he did not want his faith questioned, even after his death.

There is no way of telling how great an impact this struggle with his conscience ultimately made on his life and work. But, like the intellectual turbulence of Charles Darwin more than a hundred years later, it provides a vivid illustration of the painful dilemmas faced by pioneers in science when their work took them into areas that challenged conventional religious beliefs. The equivalent hard choices in our more secular age arise when the forces of scientific advance, in the form of genetic engineering, come up against the new faiths of environmentalism and conservation.

Sources

Since this is not an academic study, I have not included footnotes or chapter notes. Where I have quoted from a publication I have generally indicated the source in the text itself. The bibliography at the end lists these publications as well as other books, articles and pamphlets that Olga and I have found useful. Information about Fairchild's family background, including his precise relationship to his successor Stephen Bacon, was gleaned from working through records of births, marriages, deaths and wills in various record offices, although it was frustrating that the Aldbourne parish records for part of the relevant period are missing. Most of the correspondence between Bradley, Sloane and Petiver is held in the British Library, as are the lewd satires I quote in Chapter 6. Other Bradley letters are in the James Douglas Archive in the Hunterian Collection at the University of Glasgow.

Plant Names

In most cases I have tried to give up-to-date translations of names used in Fairchild's time, although the Linnaean system of naming and identifying plants did not come into use in Britain until after his death and it has been modified frequently since. Judge Gilbert Leslie (see Acknowledgements) did some painstaking work in identifying the plants mentioned in *The City Gardener*, which has saved me a great deal of time and bafflement.

The Matter of Money

When writing about money I have not attempted to translate it into today's equivalent sums because any method of doing so is unreliable, due to the fluctuating real value over the centuries of commodities, property and services, especially labour. One quite common device is to multiply all figures by a hundred to arrive at today's approximate value, but this does not work across the board. For instance, the average annual income of a farmer in the early eighteenth century was calculated at £44, and £50 was said to provide a good middle-class standard of living. Even multiplied by a hundred, these figures do not come close to present-day expectations; a factor of between five hundred and a thousand would be more realistic.

On the other hand, the cost of plants has come down in real terms because the hugely increased demand brings the benefits of mass production. Some priced stock lists from nurserymen almost contemporary with Fairchild have survived, but I have not quoted from them extensively because at this distance in time they give a distorted picture of actual values.

MICHAEL LEAPMAN

CHAPTER ONE

An Evening at Crane Court

Yes, love comes even to plants, males and females; even the hermaphrodites hold their nuptials, showing by their sexual organs which are males, which are hermaphrodites.

LINNAEUS, 1729

At 52, Thomas Fairchild, gardener and lover of the outdoors, was still a fit man, although as he had grown more prosperous he had put on flesh. At most times of the year he would have walked the two and a half miles to the City of London from his home and nursery at Hoxton, just north of Shoreditch. But this was a cold Thursday afternoon in early February 1720; there had been a frost the previous night, and the piercing north wind was enough to penetrate the heaviest cloak. Although the sky was clear, the sun was sinking fast and it would be dark soon after five. Not just that, but he was carrying a small and quite precious parcel.

Rather than tramp in his best frock coat through the muddy lanes around Smithfield, where herdsmen would be driving sheep and cattle for Friday's market, he would surely have ridden in a hackney carriage, a short stage-coach or a slow but comfortable sedan chair. For a shilling or less, he and his little package would be carried snugly from his prosperous suburb, admiring as he went the newly built Aske's almshouses and school. Passing the half-completed Charles Square, soon to be one of the most fashionable addresses in the neighbourhood, he would have been carried along Old Street to Clerkenwell, crossing stinking Fleet Ditch close to the new St Paul's Cathedral, and from there into Fleet Street, at the busy western end of the City.

It was important that he should be there in good time and good order, for this evening was to be a landmark in his professional life. After 30 years, Fairchild had gained a reputation as one of the most skilful nurserymen in England, always experimenting with new techniques to produce better flowers and fruit and to make them flourish over a longer season, and especially adept with tender or ailing plants. He was one of several in his trade known as 'curious' gardeners, in the old-fashioned sense that they displayed intense curiosity about every aspect of their craft. At a time when England was experiencing a real upsurge of interest in gardens and what grew in them, more and more people were flocking to Hoxton, then a leading centre of the trade, to gaze at the latest wonders on display at his and a clutch of neighbouring nurseries.

This evening he was going to meet some of the finest scientific minds in the country – perhaps even the great Sir Isaac Newton himself – who were to be told the results of an unprecedented experiment in botany, then a young and undeveloped science. He was heading for Crane Court, an alley on the north side of Fleet Street just east of its junction with Fetter Lane. Since 1710 it had been the headquarters of the Royal Society, the prestigious scientific institution founded in 1662, just after King Charles II's restoration. The building, which sealed the long alley at its north end, had been converted from an old house by one of the Society's founders and its most illustrious Fellow, Sir Christopher Wren – then eighty-seven and with three more years to live.

There Fairchild would have paid off his coachman or

chairmen and walked up the broad front steps, framed by iron railings and lit by a single overhead lantern. The steps narrowed as they approached the tall front door, opened for visitors by a liveried doorman whose uniform bore the Society's coat of arms in silver. That touch of obsequious ceremony – introduced by Newton, the Society's President since 1703 – would not have done anything to ease Fairchild's apprehension, and nor would the collegiate atmosphere evident in the ground-floor lobby, with groups of learned men conversing earnestly, occasionally looking towards him and wondering who exactly was this chubby outsider with the ruddy, weatherbeaten face, carrying his mysterious parcel.

The meeting-room, equally grand, was on the first floor at the back of the building, with four tall windows overlooking the garden. There was a clear division at the midpoint, where Wren had knocked two rooms into one. A low central beam was supported by two fluted columns that derived from the architect's favoured Palladian architectural style. The ceilings on either side of this divide were elaborately moulded in two similar but separate designs, with oil lamps hanging from the centre of each. Wood panelling around the walls rose about three feet from the floor, and at one end was a fireplace with moulded decoration.

Fairchild was not going to address the gathering himself but had been asked to attend by his friend Patrick Blair, a doctor, amateur botanist and a Fellow of the Society. Blair had undertaken to tell members about Fairchild's experiments, in a paper supporting his theory that plants reproduced in a way somewhat similar to animals and that, crucially, they were equipped with male and female organs. Once that fundamental principle was accepted, it would be a short step to working out how flowers could be bred deliberately, and even have their characteristics manipulated to man's design.

Today it seems odd that such a comparatively straightforward discovery was so long delayed – it was, after all, a full 30 years since Newton had published his work on the rather more complex laws of gravity. But to accept that man could so radically interfere with what God had created posed ethical and moral dilemmas in the religious climate of the early eighteenth century. Some disapproved of Fairchild's experiments and questioned whether they did not amount to blasphemy. Did they not deny the biblical account of the Creation, which credited God as the creator of all species, and had until now been taken to mean that His scheme was not subject to alteration? Was Fairchild not interfering with the prerogative of the Creator – the 'Great Author of perfection' as the Revd William Stukeley would put it in 1760, in a sermon preached in the nurseryman's memory?

There is not much doubt that Fairchild shared these misgivings. He struggled with his conscience and suffered spasms of guilt that, at the end of his life, he tried to purge through good works. His torments were comparable to those that in the following century were to plague a more celebrated pioneer in the natural sciences, Charles Darwin. No wonder, then, that Fairchild felt some trepidation as he entered the meeting-room and took his seat among the Fellows who had come to hear Dr Blair's paper.

In the event Newton was not there. Though usually an assiduous attender, he may, at 77, have been disinclined to venture into the cold night air. The meeting was chaired instead by someone not quite as distinguished but more appropriate. Sir Hans Sloane was a respected and much honoured doctor, physician to the late Queen Anne. Born in Ireland in 1660, he had been a Fellow of the Royal Society since 1685, its Secretary from 1693 to 1712 and was now its Vice-President. As Secretary he was credited with having rescued the Society from a period of financial and intellectual decline and contributed several papers to its *Philosophical Transactions*, often on bizarre subjects such as the feathers of a condor or people who ate stones.

This was a time when science was viewed as much as a freak show as a serious academic discipline, and when it was thought that knowledge could be advanced by the study of extreme phenomena. So little was known about anything, by comparison with the body of knowledge we have today, that it was impossible to be sure whether any particular odd occurrence or found object would help advance science or was merely a curiosity.

It had been like that ever since the Royal Society was founded. The diarist Samuel Pepys, admitted as a Fellow in 1665, recorded that at one of his meetings he 'saw a cat killed with the Duke of Florence's poison, and saw it proved that the oil of tobacco drawn by one of the Society do the same effect, and is judged to be the same thing with the poison both in colour and smell' – an early and vivid instance of anti-smoking propaganda. In 1682 another diarist, John Evelyn, was present when a French doctor demonstrated the first pressure cooker, making 'the hardest bones as soft as cheese'. With it, he cooked a meal of fish, beef, mutton and pigeons, which the Fellows devoured with relish.

What now seems to us as mundane, even ludicrous, commanded as much earnest attention as genuine scientific advance simply because it was impossible to judge whether or not it would turn out to have any broad significance. All this was a gift to the emerging tribe of satirists, such as the anonymous author of a periodical published in 1709 called *Useful Transactions in Philosophy and Other Sorts of Learning*. In the preface to the first issue he declared: 'It may not improperly be said at present that there is nothing in any art or science, how mean so ever it may seem at first, but that a true virtuoso, by handling it philosophically, may make of it a learned and large dissertation.'

In a later edition of the periodical, purporting to be the notes of a quizzical traveller returned from abroad, the satirist developed the theme: 'Feeding of fowl, the education and discipline of swine, the making of beds, the untying of breeches and loosening of girdles, with many other things described by this author, may seem at first to be trivial, yet contain in them great penetration of thought and depth of judgement.' Sloane himself makes an appearance, in the guise of the accident-prone Dr Van Slyboots, who remarks: 'I think it one of the most necessary things in the world for a physician when he sets up in any place, to look out for proper and convenient burying-places for his patients.' Some 200 years later, George Bernard Shaw would be making the same point in *The Doctor's Dilemma*.

Sloane had been a member of the Royal College of Physicians since 1685 and was elected its President in 1719. A wealthy man, he gave money to several London charities and was said to be considerate to his servants, including one from Africa. In 1712 he had purchased the manor of Chelsea, bordering the Thames to the west of London, which included the Society of Apothecaries' garden. In 1722, two years after meeting Fairchild at the Royal Society, Sloane would give the garden's freehold to the apothecaries, along with some money for its improvement, on condition that he remained involved in its direction. He was responsible for installing the great Philip Miller as head gardener. It is largely thanks to Sloane that the Physic Garden survives as the oldest public garden in London, with his statue by Joannes Rysbrack standing appropriately in the centre.

Sloane had developed an interest in plants and gardening when he worked in Jamaica as physician to the governor, the Duke of Albemarle. It was here, too, that he first came across the cocoa bean, and through it made perhaps his sweetest contribution to the culture of the Western world. In Jamaica he came across chocolate for the first time, observing that the local women fed it to sick children. When he returned to London in 1689 he took some beans with him and experimented by mixing their powder with milk. The result was so palatable that he sold the recipe to a London grocer, whose successors sold it on to the Cadbury brothers, whose milk chocolate was to conquer the world.

Because early eighteenth-century medicine was based to a large extent on herbal remedies, many doctors became involved in botany. When Sloane returned from Jamaica he brought with him, as well as the cocoa beans, samples of some eight hundred plants that grew there. He published a catalogue of them seven years afterwards, eventually expanding the work into an account of his journey and a full-scale natural history of Jamaica. But as his biographer E. St John Brooks has observed, 'like many fashionable doctors he was a courtier rather than a scientist'. He hosted a weekly dinner party at his Bloomsbury house and invitations were highly coveted: although the fare was not extravagant, he would sometimes serve game sent to him by landowning patients. Later he was to be a prime mover in the establishment of the British Museum.

Fairchild was the only man in the room on that cold February night who was not a Fellow of the Royal Society. He was never to become one, probably because his education had been practical rather than intellectual. After the minutes of the previous meeting had been read and approved, the Fellows had to give their formal consent to his attendance. That done, Blair stood up to present his paper. He had just published a book of botanical essays, mainly concerned with plant reproduction, and he had offered to mark its appearance by talking to the Royal Society about his conclusions. He was unashamed about promoting his work, going so far as to give specific page references for his theories about sex in plants and the circulation of sap that were confirmed by the experiments he was describing.

The first of them had been carried out by Thomas Knowlton, then a young gardener at Offley Place near Hitchin in Hertfordshire, later a close friend of Fairchild's and ultimately one of the most influential horticulturalists of the eighteenth century. Knowlton, who was not present that evening, had used two different methods of sowing wheat. He put some in rows, sowing each grain individually, and the other batch he scattered in drills 'promiscuously', as Blair put it. 'That which was sown singly shed its dust [pollen] before the female embryo began to appear,' he reported. Hardly any of that batch ripened, and the yield was minimal. The other seeds, however, produced plentifully. This, said Blair, 'confirms that the union of male and female flowers is necessary to fructification'.

Then he came to Fairchild's experiment. As Blair described it, the Hoxton nurseryman had found in his garden a plant 'of a middle nature between a sweet william and a carnation', at a spot where seeds of the two flowers had been scattered accidentally – close enough for the pollen of one to enter the stigma of the other. Responding to Blair's cue, Fairchild opened the package he had been cradling so carefully and produced a specimen of the unique flower, pressed and preserved, and passed it around the gathering.

Blair added that Knowlton had found a very similar flower at Offley, apparently a cross between a sweet william and a China pink. Neither of these freaks had produced any viable seed. They were 'barren like the mule [a cross between a horse and a donkey] or other mongrel animals which are generated from different species'. Over the next hundred years the Hoxton flower, reproduced from cuttings, became quite popular among gardeners and was known as Fairchild's mule. (The word 'hybrid', from the Latin term for the product of a union between a tame pig and a wild boar, did not come to be used of plants until much later.)

Blair's account of Fairchild's experiment differs in one important detail from that given by his friend Richard Bradley, a prolific and controversial horticultural writer, also a Fellow of the Royal Society, who had made the first known reference to the mule four years earlier in his book *New Improvements of Planting and Gardening, both Philosophical and Practical*. He wrote that the bastard flower was not the result of an accidental discovery but that Fairchild, famed for his well-developed curiosity about his craft, had deliberately impregnated a carnation with the farina (pollen) of a sweet William.

How can this discrepancy be explained? So widespread was the conviction that it was wrong to seek to impose man's will on God's creations that Fairchild may have asked Blair to let the grandees of the Royal Society believe that his discovery had been made by chance rather than design. Hybridisation, after all, is a precursor to genetic engineering, which provokes the same inhibitions and objections today. Both involve crossbreeding to produce new kinds of plants that would not occur if nature were left to take its course. The eighteenth-century opponents of such experiments deployed essentially the same argument as those who decry genetic modification – that the consequences of tampering with nature are unknown, and that the risk involved is therefore unacceptable.

A second theory about the conflicting accounts of the discovery was put forward by Conway Zirkle in his 1935 book *The Beginnings of Plant Hybridization*. He believed that there were two separate incidents, that Fairchild came across the hybrid accidentally and then, with his enquiring cast of mind, set out to produce one deliberately: 'It would not be in keeping with Fairchild's character for him to find a natural hybrid and experiment no further.' Certainly he bred many other examples of the mule from cuttings, or from further original crosses.

Whatever the exact circumstances of the discovery, the effect on Fairchild when he first saw that alien flower must have been shattering. The era of the plant-hunters, those adventurous men who undertook dangerous voyages to the ends of the earth to discover new floral wonders, had just begun in earnest and was not to reach its zenith until the following century. Yet here was a brand-new flower, just as exotic in its own way as any brought back from overseas, flowering in Fairchild's modest nursery just north of London, bred by a man who, gifted though he was in the mysteries of the flowerbed, could by no stretch of the imagination describe himself as a scientist.

In the circumstances, Fairchild may have found Blair's

deadpan account of the event a mite undercharged. Quite quickly, the learned doctor moved on to describe to the Society's Fellows some experiments relating to the flow of sap in trees and plants, then a popular field of study among botanists. Many physicians believed it bore comparison with the circulation of blood in humans and animals, discovered by William Harvey a century earlier.

First Blair showed his audience two small branches of a pear tree, one of them 'circumcised' and the other not. Circumcising a tree meant stripping a ring of bark from around a branch to stimulate the circulation of sap. He reported that the branch so treated produced plenty of flower buds, while the untreated branch had only leaf buds. Then he referred to an experiment in grafting that Fairchild had carried out some years earlier, when a cutting from a fertile pear tree had been grafted on to the rootstock of an infertile one, resulting in a tree more fruitful than either – a technique still used today.

Fairchild had recognised early on that understanding the mechanics of sap circulation was a key element in successful grafting: that to be a real master of any craft you had to understand not just how a procedure worked but why. Blair ended his presentation by describing the grafting of an evergreen oak on to a deciduous stock, and some experiments with honeysuckle and vines. These, too, had probably been carried out by Fairchild, whose skill with grapes was renowned.

When Blair spoke proprietorially of *his* theories about plant reproduction, he was taking liberties with the literal truth, for he was merely giving support to a notion that had been gaining currency in Europe for the last half-century. Compared with our northern European neighbours, the British had lagged behind in both the practice and theory of gardening. In part this was because the Civil War of the 1640s had interrupted domestic life for many of the landed noblemen who would, in peaceful times, have become patrons of the emerging skill. Then in the 1660s a series of plague epidemics hindered the return to everyday pursuits. The work of the great Flemish botanist Charles de l'Ecluse (Clusius), whose *Rarorium Plantarium Historia* was published in 1601, had been enormously influential on the European mainland but had less of an impact in England, where botany was a neglected science and where John Gerard's *Herbal*, appearing four years earlier, was sloppy and amateurish by comparison. Better herbals appeared in the seventeenth century, notably Nicholas Culpeper's, but these were essentially lists of plants believed to be effective remedies for disease and injury – guides to medicine and astrology rather than botany.

Clusius went plant-hunting in Asia and came to Britain twice, in 1571 and 1581, when he gleaned information from Sir Francis Drake about the plant life of the Americas. As superintendent of the Royal Gardens in Vienna and head of the earliest botanic garden in the Netherlands, he was responsible for the introduction of many tulips; but he was dead by the time of the speculative tulip mania that swept the Netherlands in the 1630s, when unbelievable prices were paid not just for the tulip bulbs themselves but also for contracts for their supply, and fortunes were lost. The tulip madness never captured the British to the same extent, despite heartbreaking stories of servants cooking and eating their masters' precious but unfamiliar bulbs, assuming them to be a kind of onion. (Clusius himself tried eating them, as did the British botanist John Parkinson – but strictly for the purposes of research.)

Today eating tulip bulbs would be an aberration; but until the seventeenth century, botanists were concerned chiefly with the food value and curative qualities of plants rather than how to breed new and decorative forms. The science of botany was a strictly practical one, aimed at determining how plants could be made useful to humans, rather than analysing their structure and reproductive systems. Gardening was still primarily a branch of medicine, and in the seventeenth and eighteenth centuries many botanists were, like Sloane, physicians who studied not just plants but other branches of natural history – zoology and even anthropology – as a hobby, albeit one closely related to their work. Thus they would examine the anatomy of a lily or an elephant with the same curiosity as they would the anatomy of a dead person. The distinction between scientific disciplines was less clear cut than it is today.

In the early herbals, the shape and colour of flowers were noted chiefly for the purpose of plant identification, not appreciated for their intrinsic beauty. Contemporary descriptions of the gardens of Tudor houses concentrated on their design features – elaborate parterres, fountains and statuary – rather than their floral displays. Sir Thomas More, in his book *Utopia*, published in 1516, described an ideal townscape where each house had a well laid-out garden behind it, containing vines, fruits, herbs and – last and probably least – flowers:

> I never saw thing more fruitful or better trimmed in any place. Their study and diligence herein cometh not only of pleasure but also of certain strife and contention that is between street and street concerning the trimming, husbanding and furnishing of their gardens, every man for his own part.

(Garden envy, then, is of ancient origin.) Later, the many references to flowers in Shakespeare's plays and sonnets, and their increasing incidence in seventeenth-century paintings and decorative motifs, show that they were starting to become appreciated in themselves.

The restoration of the monarchy in 1660, after 11 years of stern rule by the Puritans, sparked renewed enthusiasm for the decorative arts, and gardening was by now recognised as one of them. All the same, in 1665 John Rea, a noted nurseryman, could still write, in his book *Flora*: 'fair houses are more frequent than fine gardens', though he indicated that flowers were starting to be seen as at least as important as architectural features: 'A choice collection of living beauties, rare plants, flowers and fruit are indeed the wealth, glory and delight of a garden and the most absolute indications of the owner's ingenuity.' He described gardening as 'this lovely recreation' and listed the most commonly grown species: auriculas, primulas, campions, violets, wallflowers and gillyflowers (known today as carnations). Even in those early times, flowers were subject to the vagaries of fashion. Rea wrote that many plants listed by John Parkinson in his *Paradisi in Sole, Paradisus Terrestris* in 1629 had 'by time grown stale and for unworthiness turned out of every good garden'.

When William and Mary arrived from Holland in 1688 to assume the British throne, they fostered an interest in all things Dutch, including gardening, one of the king's particular passions. They introduced an immense number of exotic plants to their gardens and hothouses at Hampton Court Palace, many coming from South Africa, first settled by the Dutch in 1652. They included yuccas, cacti, palms, aloes and even coffee plants. In the Royal Horticultural Society's magazine *The Garden* in May 1999, Mark Griffiths wrote: 'Amassing these plants was no less important for the joint monarchs than collecting paintings or sculpture might be for other great patrons. Exotics were, in effect, living art works.' Stephen Switzer, a contemporary gardener, seedsman and writer, tells us that Queen Mary was just as enthusiastic as her husband, and particularly skilled with exotics.

The big growth of seventeenth-century gardening was in the country outside London. In 1677 John Worlidge, in his *Systema Horticulturae*, wrote that there was 'scarce a cottage in most parts of the southern parts of England but hath its proportionable garden, so great a delight do most men take in it'. Paradoxically, though, most of the nurseries were in the London area, and towards the end of the century domestic horticulture finally began to spread to the metropolis. By 1691 the essayist John Aubrey could write that there was 'ten times as much gardening about London as there was in 1660'. Daniel Defoe, in *A Tour through the Whole Island of Great Britain*, noted that a few years after the accession of William and Mary 'fine gardens and fine houses began to grow up in every quarter', especially in Middlesex and Surrey, the counties whose common border was the Thames flowing through London.

The owners of these new gardens sought novelty above all else. They vied with each other to possess the very latest and most spectacular varieties of trees and plants – a sure way of earning prestige among their peers. Yet they understood little about how plants grew as they did. Their ignorance may have made the wonders of nature seem yet more wondrous – but at the same time it severely limited their horticultural options.

There is evidence that the Chinese had been hybridising roses and camellias for centuries, but the technique had not spread to Europe. Because no method then existed of deliberately breeding 'new' plants, the range of flowers grown was inevitably small. Innovation could occur in three ways. The first was by accident, when a naturally occurring hybrid that differed from both its parents was discovered in someone's garden or in the wild. By taking cuttings from the 'freak' plant, replicas could be grown. Zirkle says this happened as far back as Neolithic times, when farmers growing cereals would notice a new hybrid that produced better crops and would breed from it vegetatively.

The second method was by selection, particularly for colour. This was (and still is) done by uprooting plants of the less favoured colour before they had a chance to seed, leaving only those of better colour to produce seed for the following year. The third way was plant-hunting: voyaging to distant parts of the world in search of new flowers to introduce into England.

These expeditions were financed by doctors (including Sloane himself), botanists and other scientists, and later by nurserymen and seedsmen, as well as by the owners of great gardens seeking novelties to amaze their friends. That is why the voyagers brought back not only plants but a treasure-trove of other artefacts – animal bones, totem poles, precious stones – to illuminate the culture of distant lands. The first notable plant-hunter was John Tradescant the Elder, gardener to the nobility and the royal family, who sailed to America and elsewhere in the early years of the seventeenth century and brought back many new flowering plants, including some that are still named after him, such as *Tradescantia virginiensis*. He also collected quantities of more durable 'curiosities' and displayed them at The Ark, his house in Lambeth, South London: some are still to be seen at the Ashmolean

Museum in Oxford. His son, John Tradescant the Younger, carried on his father's pioneering work. Fairchild, too, had plant-hunters looking out for novelties for his nursery, among them the celebrated Mark Catesby, who made two long trips to America and the West Indies.

In the early days of gardening, nurseries fulfilled a dual role. They were not simply places at which to buy new plants raised by the nurseryman; they also acted as horticultural boarding houses for plants that could not be kept year-round in their owners' gardens, and as intensive-care units for those that were especially delicate or damaged. The booty brought back by the plant-hunters had encouraged a growing demand for tender plants that would not survive outdoors in a British winter; yet only the very biggest estates could afford expensive 'stoves', or hothouses, to keep them in. So some collectors would send their precious specimens to specialists such as Fairchild, to be nursed through the season of greatest peril. That is why nurserymen of the time, advertising their services, stressed the excellence of their heating systems as much as their horticultural and botanical skills.

While seventeenth-century gardeners were becoming increasingly adept at making things grow, there was no body of scientific knowledge to tell them precisely why they were succeeding. Obviously, they knew that plants grew from seeds, but exactly how the seeds were fertilised was a mystery to them. They were baffled when seed taken from, say, a red flower would produce blooms in a variety of colours. These mysteries challenged and intrigued the early botanists, chief among them John Ray, whose work on the identification of plants, undertaken on exhaustive tours of the country, prepared the groundwork for the Linnaean system, the first universally accepted system of plant classification that is still, with modifications, in use today.

The son of a blacksmith, Ray was born in Braintree, Essex, in 1628. He went to Cambridge University as a student and stayed on to lecture in Greek, mathematics and the humanities. In 1660 he was ordained a priest. He published a catalogue of English plants in 1670 and, 16 years later, his renowned *Historia Plantarum*. An innovative scholar, he is believed to have been the first to use the word 'petal' to describe the coloured leaves of flowers that surround the seed head. Sloane, although a younger man than Ray, was a friend and admirer.

On 17 December 1674 Ray read two papers to the Royal Society that give us a good idea of the state of knowledge about plant reproduction at the time. His original manuscripts, in his neat but cramped hand, are preserved in the Society's archives. The first of the papers was about the structure of seeds and their reproductive mechanism. He began by explaining why the biggest seeds do not necessarily produce the biggest plants. There were, he argued, various different types of seed, and some effectively contained whole, miniature plants within them. He believed that the key to their growth was how they took up the nourishment they needed to grow to full size. He had at first thought that the seed pods themselves contained food for the kernel to ingest, but he had now changed his mind and believed that all the nourishment came from the earth. He compared it with the birth process of animals in the womb:

> For as the seed of a plant when ripe falls to the ground and through lying loose does [as I saw] first receive its nourishment by the pores of its legumens [seed pods]

> and afterwards strikes root into the earth, so likewise the seed and egg of a viviparous animal, when ripened as it were by the male, drops off one of the ovaria into the womb where it lives for a while loose or free, without any adhesion to or connexion with the womb, drawing its nourishment through its involving membranes, then striking root into the womb.

By likening plants to animals, Ray was at least getting close to accepting the idea of sexual differences among them, but he was not quite there. In using the image of female eggs being 'ripened' by male animals, he suggested that the sun fertilises plant seeds – in other words that conception does not occur until after the seed has left the flower. This is a logical conclusion if the idea of plants with male and female organs seems too preposterous to consider. Only 50 years earlier, Francis Bacon, in his *Sylva Sylvarum*, had predicted the possibilities of producing desirable new plants if their method of reproduction could be discovered, but he declared decisively that 'generation by copulation certainly extendeth not to plants' – a view then widely accepted.

Ray's second paper was on the specific differences between plants and how they occurred. It would be almost a hundred years before Linnaeus (Carl von Linné), the leading botanist of his time, carried out his definitive work on dividing plants into species, and there was still confusion about how closely one was related to the other. Earlier botanical writers such as John Gerard and John Parkinson often classed as different species plants that we now accept to be varieties of the same species. Ray maintained that 'accidents' of various kinds could cause considerable variations within species – differences in size, variable shape and colour of roots (such as in turnips and carrots), variegated leaves and differences in flower colour. In his garden, he disclosed, the seeds of a yellow-flowered mullein (verbascum) had produced white flowers. Such differences were not sufficient, in his view, to amount to a 'specific distinction'. He explained:

> Diversity of colour in the flower or taste in the fruit is no better a note of specific difference in plants than the like varieties of hair or skin or taste of flesh in animals: so that one may with as good reason admit a Blackamoor and European to be two species of man, or a black cow and a white to be two sorts of a kind, as two plants differing only in colour of flower to be specifically distinct.

He pointed out that gardeners could affect the colour of a flower by watering it with a solution of the desired colour, and that if seeds were planted in an unusually rich soil they tended to produce more than the normal quota of double flowers. This meant that many aspects of a flower's appearance were subject to environmental influences, rather than being attributes inherited from its parents. To ensure that a plant reproduced itself exactly, it was necessary to take a cutting from its roots or leaf stems rather than to grow it from seed. By this method, variations worth growing in gardens could be reproduced ad infinitum.

Ray ended his talk with a sentiment that gave spiritual correctness to his scientific findings. He argued that while the number of variations within species was infinite, it was impossible to devise entirely new categories of plants, 'the number of species being in nature certain and determinate, as is acknowledged by philosophers and might be proved also by divine authority, God having finished his works of creation, that is consummated the number of species, in six days'. Within forty years Fairchild's experiments in hybridisation had effectively given the lie to that fervently held belief – although he was as reluctant as anyone else to face the true significance of what he had done.

Just who made the theoretical breakthrough concerning sex in plants has never been properly established. In an essay in 1760 Linnaeus wrote: 'To say exactly who first came upon the sex of plants would be a thing of great difficulty, and no use.' It had been known for centuries that date palms are either male or female and that the female plants have to be fertilised by the male: Babylonian monuments from about 650 BC show the process being done by hand. But for centuries it was thought that this was peculiar to the date palm, since no other plant was known to behave similarly; and it was thought relevant only in terms of production of the fruit, not of propagation of the plant.

Ray would have been aware of a work published in Italy in 1671, three years before his Royal Society lecture. Marcello Malpighi, a doctor from Bologna, wrote *Anatomia Plantarum*, the first recorded publication that mentions the possibility of sexual reproduction in plants in general. The paper was translated and read to the Society. He would also have heard the theories of Nehemiah Grew, a doctor and botanist who had given a number of talks to the Society about the structure of plants in which he was clearly moving towards the truth about their method of reproduction and its relation to that of animals. He wrote that plants and animals 'came at first out of the same Hand, and were the contrivances of the same Wisdom'. The analogy with animals could never be absolute, because there is no animal equivalent to vegetative reproduction through leaf and stem cuttings – or at least there was not until 300 years later, when the possibilities of animal cloning came to be realised.

In a lecture to the Royal Society in 1676 and in *The Anatomy of Plants* in 1682, Grew aired his theories with increasing conviction and in greater detail, noting specifically that the stamen was the male part of the flower head and that 'powder . . . like male sperm serves to fecundate the seed; and therefore that most plants partake of both sexes'. Ray, in his *Historia Plantarum* four years later, accepted the probable truth of these notions, which the invention of the microscope earlier that century had made easier to confirm.

In 1694 Rudolf Camerarius, of the University of Tübingen in Germany, made the strongest scientific case yet for sexual differentiation in plants. He is regarded as the first to prove by experiment rather than deduction that pollen is necessary for fertilisation and that only male flowers produce it, whereas female flowers bear seed. He noted that some plants, such as maize, are bisexual, carrying both male and female flowers in the same head. He wrote:

> They behave indeed to each other as male and female, and are otherwise not different from one another. They are thus distinguished with respect to sex, and this is not to be understood as it is ordinarily done, as a sort of comparison, analogy or figure of speech, but it is to be taken actually and literally as such.

Bradley's 1717 book demonstrates how far this theory had gained ground in the last few decades. He says that the first hint he had of the sexual nature of plants was given to him some years earlier by Robert Balle, another Fellow of the Royal Society, 'who had this notion for above 30 years,

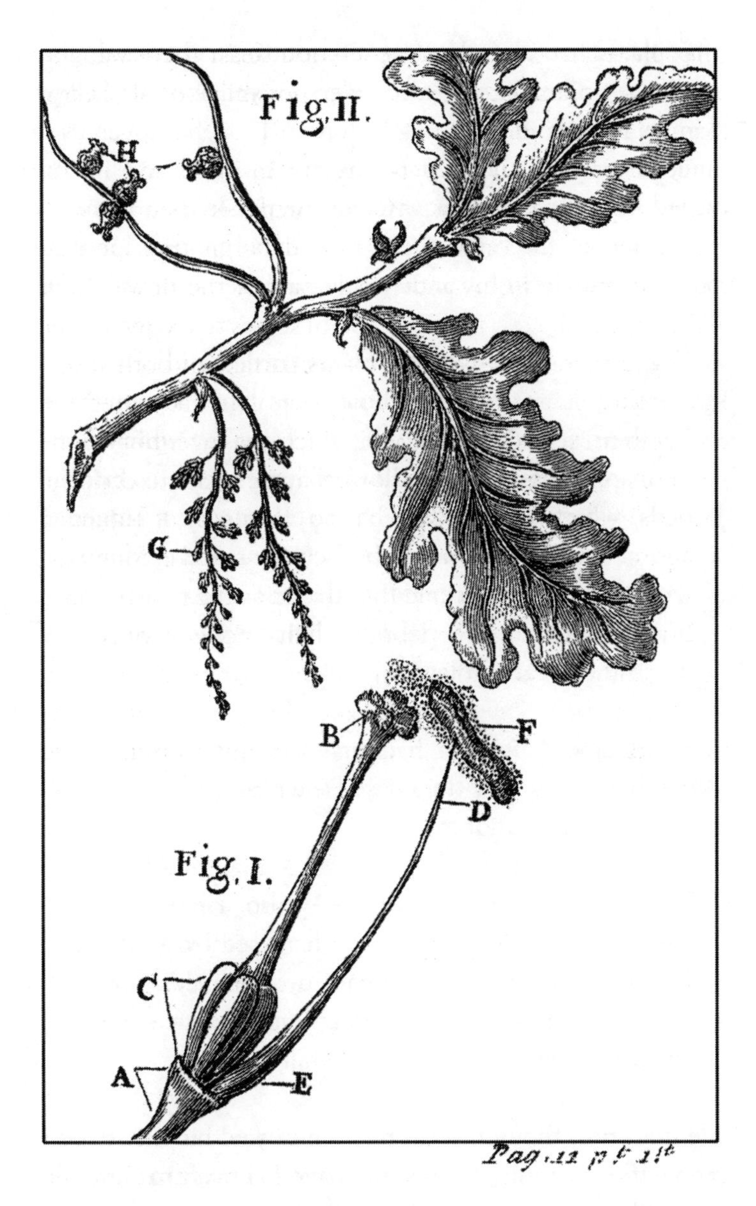

The sexual structure of plants and flowers illustrated in Bradley's *New Improvements of Planting and Gardening* (1726)

that plants had a mode of generation somewhat analogous to that of animals'. He also refers to yet another Fellow, Samuel Moreland, who in a lecture in 1703 'has given us to understand how the dust of the apices in flowers is conveyed into the uterus or vasculum seminalis of a plant, by which means the seeds therein are impregnated'. Linnaeus put it spiritedly in his undergraduate dissertation in 1729:

> Yes, love comes even to plants, males and females; even the hermaphrodites hold their nuptials, showing by their sexual organs which are males, which are hermaphrodites . . . The flowers' leaves serve as bridal beds, which the Creator has so gloriously arranged, adorned with such noble bed curtains and perfumed with so many soft scents that the bridegroom with his bride might there celebrate their nuptials with so much the greater solemnity.

It was Camerarius who had first thought those theories through and saw how they raised the possibility of what we now call hybridisation:

> The difficult question, which is also a new one, is whether a female plant can be fertilised by a male of another kind, the female hemp by the male hops . . . and whether, and in what degree altered, a seedling will arise therefrom.

He does not, though, seem to have carried out any experiments to prove this. That would have to be left to Fairchild, the skilled gardener, a man wedded to practice rather than theory.

Records of Royal Society meetings contain details only of the papers and not of any discussion that may have followed them. They do not tell us how the Fellows reacted to Blair's revelation of Fairchild's experiment, or whether they recognised how significant it would be for the future of botany and horticulture. Given that it would be many years before plant hybridisation became common, it seems likely that they did not appreciate the historic nature of the occasion. The record shows merely that the Fellows thanked Blair for his paper and settled down to hear Lord Percival read a letter from his brother in Dublin describing a meteor that had been seen there the previous month. Afterwards Fairchild showed them another curiosity that he had brought with him, a piece of wood in which a chrysalis had 'formed a case or coffin to lodge in till its transmutation into a moth'. He was thanked for that also, and the meeting ended with a description of 'a monstrous birth' of Siamese twins three weeks earlier.

It had been a quite typical evening at Crane Court, where the most inquisitive minds of the age met regularly to exchange information on natural phenomena that they or others had observed, seeking from them to identify laws that would give a better understanding of how the universe worked. From today's vantage point we can judge that some of their conclusions were skewed or plain wrong, but they were operating at the interface of the known and the unknown, straining constantly to expand the boundaries of science, and at the same time trying not to fall foul of the prevailing theology. It was a delicate balancing act; but without their agile and questioning minds, our twenty-first-century world would be different and poorer.

We have it on the authority some years earlier of the diarists Samuel Pepys and John Evelyn, both Fellows of the

Royal Society, that after their meetings the members would repair to one of the many City inns or coffee houses for supper. For several years they used the best room at the Crown in Threadneedle Street, convenient for Gresham College, where the Society met until 1710, but not within comfortable walking distance of Crane Court, especially in midwinter. Whatever the venue, Fairchild would probably have been invited to it by Blair or Bradley to engage in some solid garden talk before making his way home. Sloane might have joined in too, assuming he did not have to dash away to minister to one of his aristocratic patients.

CHAPTER TWO

Roots

The gifts of nature are much more valuable than those of original birth and fortune, or even learning itself.

STEPHEN SWITZER, 1715

At the corner of Pitfield and Minton Streets in Hoxton today lies a small, nondescript public park, the entrance flanked by holly bushes, with a few hybrid tea roses struggling to flower in a bed just to its right. Beyond the fence are playing fields and a view of a 1970s leisure centre. South of the park, across Buckland Street, is St John's Estate, a standard postwar development of high-rise council flats built around a shaded open space where trees dot the lawn. In the centre is a small bed containing the kind of shrubs that can survive in such inhospitable surroundings – pyracantha, laurel and the like – and beyond it a tarmac play area.

Improbable as it now seems, this drab urban landscape was the site of Fairchild's nursery. His book *The City Gardener*, published in 1722, has as its frontispiece a drawing of a garden that might represent his own but is more likely an idealised version of a well-tended plot. It is hard to imagine how he would have raised the capital to create such an impressive place for himself, and there is no sign of a sales area or of planting on a commercial scale. It is elegantly laid out with broad paths, where two gentlemen in wigs and frock coats, carrying swords, are walking between tubs containing large and quite exotic flowering shrubs. Blanche Henrey, the garden historian and former

Frontispiece from Thomas Fairchild's *The City Gardener* (*Royal Horticultural Society, Lindley Library*)

archivist at the Natural History Museum, has identified the shrubs as an agave, a banana, a dwarf palm and a cactus – all tender plants that would have had to be moved into heated quarters during the winter.

A dog with an unusually large head prances in front of the men while two gardeners wearing smocks and shapeless hats, one with a hoe and the other a spade, are working on formal flowerbeds on either side of the path. Beyond are two brick structures with sloping roofs, almost certainly 'hotbeds' for tender plants, including the newly fashionable pineapples. They conform quite closely to Bradley's description of a hotbed in the Duke of Chandos's garden at Canons in Middlesex – a brick-lined pit, some five feet deep and seven and a half feet wide, covered with glass. Hot dung was piled into it, then covered with a thick layer of tanner's bark that would keep the heat in from autumn to spring. (Leather was tanned by being steeped in a solution of bark, generally oak. The spent bark could be used as a garden mulch.) Behind the beds is a straight line of regularly spaced trees with neatly rounded crowns, possibly marking the boundary of the property.

By the late eighteenth century Fairchild's nursery had been built over, and in Victorian times a workhouse was constructed here. No trace of the nursery remains, and no plaque marks the spot. Yet were it not for the work he carried out here, the best-loved flowers in our gardens might be less handsome, their scent less beguiling. Those hybrid tea roses near the park entrance would not be half as showy. The range of varieties available would be severely restricted; and gardening might never have become the national obsession, nor the multi-million-pound industry, that it is today. Fairchild's mule was the first of many thousands of crosses between flowers that have produced such a huge choice for latter-day gardeners.

He was a significant contributor to the *Catalogus Plantarum*, an ambitious plant directory published by the Society of Gardeners in 1730, a year after his death, and the book's foreword celebrates the revolution he helped foment: 'Within the space of 50 years the practical part of gardening began to rouse out of the long lethargy it had lain in.' Dr James Douglas, a contemporary physician and garden enthusiast wrote of Fairchild:

> He is so deservedly famous for his great skill in all the parts of gardening, that to him we are obliged for most of the considerable improvements that have been made in that delightful art for these several years past.

And Richard Pulteney, a renowned botanist born the year after Fairchild's death, described him as one of the three or four most influential gardeners of his day.

Fairchild was born in 1667 in Aldbourne, a prosperous Wiltshire farming village between Swindon and Hungerford which had played an important part in the Civil War a few years earlier: the Royalist armies inflicted several small defeats on the Parliamentarians between 1643 and 1645, and Cromwell's troops did some damage to St Michael's Church when they were quartered in it. The village lies deep in a sheltered valley south of the White Horse Hills, amid countryside so lush that the diarist John Aubrey, in his *Natural History of Wiltshire*, wrote that Aldbourne's rabbits were 'the best, sweetest and fattest in England', and Bradley reported in 1724 that, according to the Master of the Warren, their number had increased from 8,000 to 24,000 in a single year. Bradley also claimed fine qualities for the village's dairy produce: 'The cheese made from the Aldbourne cows is much richer and fatter than that from the cows in the vale.' (He was referring to the Vale of Kennet, a few miles south.)

The rabbits caused trouble for one of Fairchild's ancestors, John Fairchild, possibly his great-grandfather, who was convicted in 1610 of 'hunting and killing coneys [rabbits]' belonging to the Earl of Pembroke in Aldbourne. There were ten separate counts involving some 30 rabbits, which he snared by digging into their warrens and setting dogs on them. On one occasion he hid three rabbits under his cloak and, when challenged, said they were sandstones. He was committed to prison but the judge ruled that since he was 'a very poor fellow' he could be released if he entered a bond of either four or forty pounds (the record is indistinct but the lower sum seems more likely) and undertook to be of good behaviour.

The Fairchilds were quite a large clan, but there were more of them in the vicinity of Lambourn, just across the county border in Berkshire, than in Aldbourne. A half-timbered and thatched cottage called Fairchild Cottage, parts dating from the sixteenth century, stands next to the church at Eastbury, on Lambourn's outskirts, with phlox and loosestrife blowing in the front garden. A nineteenth-century John Fairchild is buried in the churchyard.

Thomas Fairchild's father was also called John, one of six of his parents' children who reached adulthood. By 1665 he appears to have been farming a small acreage of copyhold land in Aldbourne. The copyhold system, abolished only in 1922, was in effect a bridge between the feudal system and modern land tenure arrangements. Although the land belonged to the lord of the manor, the copyholder had rights almost amounting to a freehold but needed permission from the landowner to buy, sell, sublet or inherit the property. A fee was charged for these transactions, which were written in the court roll and a copy of the entry given to the tenant – hence the name of the system.

In June 1665 John Fairchild married Ann Shepherd, a widow, whose maiden name was Ann Butt. She was born at Minchinhampton in Gloucestershire in 1636, but soon afterwards the family moved to Lambourn Woodlands, not far from Aldbourne. Ann's first husband, Thomas Shepherd, had died the previous year when their daughter, also named Ann, was three. The records suggest that little Ann died two months after her mother's second marriage.

Thomas Fairchild was the only surviving son of the second marriage: the first child, John, died in September 1666 when just five months old. Thomas would have been baptised in St Michael's Church, which stood then, as it does now, on a slope overlooking the village green. Although it underwent some Victorianisation in the 1860s by the architect William Butterfield, Fairchild would recognise the church were he to return today, with its fifteenth-century tower standing at the west end of the largely thirteenth-century nave and chancel, incorporating some features from an earlier Norman church on the site. A dog's head carved in the eleventh century has been placed between the arches near the octagonal font, which is old enough for Fairchild to have been baptised in.

In 1668, when he was a year old, his father died, widowing his mother for the second time. This was sad but not at all unusual for an era when more than half of marriages lasted less than ten years because of the death of one of the partners. As Peter Earle writes in *A City Full of*

People: 'Many of the surviving partners remarried, often several times, producing a pattern similar to that created by divorce today, with a confusion of "mixed" families containing step-parents and half-siblings' – precisely the position of Thomas Fairchild.

John Fairchild left an estate worth £54.12s., most of it in the form of livestock and corn in his barn. The sum represents slightly more than a year's income for the average farmer at that time, estimated at £44. He bequeathed all but £20 to his widow. His year-old son Thomas was to receive the £20, half of it to go to his mother so that 'his life shall be put into the billing', a reference to the fee required for having the copyhold land on which John Fairchild farmed transferred into his son's name. The other £10 was to be paid to Thomas when he became 21.

Since the Aldbourne court rolls for that period do not appear to have survived, it is impossible to say whether Ann Fairchild complied with her husband's wishes and staked young Thomas's claim to the copyhold. If she did, he may have been able to sell his rights when it became clear that he did not want to farm his father's land, and this could have been where he acquired the capital he would have needed to start his nursery.

The inventory of property attached to John Fairchild's will indicates that the income from his farm was only enough to keep his wife and child in the basic necessities of life. The value of his clothing, together with the money in his purse, was put at £5. There were two tables, a barrel, three kivers (large tubs), a butter churn, benches, one chair and 'other lumber', plus a small amount of cooking equipment and four pewter dishes. Any pottery cups and dishes were presumably of too little value to be listed. The main bed was equipped with just one set of bedding: two sheets, two blankets, two pillows and a bolster.

In June 1670, two years after John Fairchild's death, his widow Ann took a third husband, John Bacon. There had been Bacons in Aldbourne since at least the previous century, and a 1649 survey records that a Stephen Bacon held a copyhold on a cottage attached to the rectory. John and Ann, though, were married in Highworth, north of Swindon, where John, who apparently could not write, made his mark, a reverse capital B, on the marriage bond.

The children of this marriage – Stephen, John and Ann – were born in Aldbourne, and it is likely that Thomas (who was able to write) lived there with them, perhaps going to school in a room over the church porch. Frank West, in his history of the church published in 1987, records that this school – the only one in the village – was closed in the eighteenth century through lack of support, but it is unclear how long it had been in operation. The floor of the room has since been removed and the porch is now one high-ceilinged space, but just to the east of it, inside the church, are remnants of the narrow staircase that led up to the schoolroom.

By the time he was 15, Thomas had moved to the City of London, possibly living near St Giles in Cripplegate, where in 1682 he took up a 7-year apprenticeship to a clothworker named Jeremiah Seamer. He was the first Fairchild to join the Clothworkers' Company – but not the last, for a few years later John Fairchild of Aldbourne, Thomas's cousin, took up an apprenticeship with Samuel Major and became a freeman of the guild in 1696. Aldbourne had a small clothmaking industry, which may explain the connection. Cloth finishing was a comparatively new trade, almost entirely concentrated in London.

In her book *Restoration London*, Liza Picard writes that by the end of the seventeenth century it represented the largest national source of income.

The companies or guilds controlled nearly all of London's skilled trades. To set up in business in one of them it was necessary to become a freeman of the company and of the City. This could be done in one of three ways – by patrimony, if your father was a freeman of the guild; by servitude, which meant serving as an apprentice for seven years; or by making a much higher payment than was involved in those two methods and becoming a freeman by redemption.

Normally, Thomas Fairchild would have expected to become a freeman of the Clothworkers' Company by servitude at the end of the apprenticeship, but in fact he did not take this step until 1704, 15 years after he would have qualified. The reason for this delay was that he had decided, almost as soon as he began his training, that indoor work was not for him and that he wanted, like his father, to be engaged with the soil. When he began his gardening career can be established with some certainty from *The City Gardener*. In it he says that he has been a gardener 'for about forty years', which would mean that he took it up at about the same time as he began his apprenticeship – perhaps moonlighting or perhaps doing garden work for Seamer. One way or another he completed his apprenticeship as a clothworker – otherwise he could not have become a freeman of the guild by servitude – but very soon afterwards he went full time into the burgeoning nursery trade, where he would make his name.

In about 1690, just after his apprenticeship had run its course, he took a position in the Hoxton garden that became his nursery, eventually running it until his death in

1729. Again, we can put a date on his arrival from *The City Gardener*, where he wrote that he had been working the same London land 'upwards of thirty years'. It is unlikely that he would have set up on his own straight away with scarcely any relevant experience. More probably, he joined the nursery as an employee and took it over a few years later: the rate books show him as being the occupant at least by 1703. No records survive of the nursery's previous history, of when or how he joined it or of how he could afford the lease. The £10 bequeathed to him by his father, which he was due to inherit when he became 21 in 1688, would scarcely have been enough by itself, seeing that the records show his annual rates bill to have averaged about £1.

That the son of a Wiltshire yeoman farmer could set himself up in business and eventually consort with London's intellectual and social elite was a symptom of the new fluidity in English society in the late seventeenth century, in the aftermath of the Civil War that had disrupted the old certainties and made it easier to breach class divisions. Gardening, the new enthusiasm of the upper and middle classes, provided a perfect context for the creation of a breed of tradesmen with specialised knowledge that their clients and patrons needed and valued. Nurserymen and gardeners could thus command more respect than less skilled people such as shopkeepers, labourers and domestic servants. They were among the first professional men, in the sense of practitioners who had gained their expertise outside formal institutions of learning. (It is interesting that the word 'professional' did not take on that meaning until the early nineteenth century.)

Fairchild's was one of at least half a dozen nurseries in Hoxton, then fast becoming the principal horticultural district in London. The late seventeenth century had seen substantial residential developments in the area immediately to the north of the City. In 1724 Daniel Defoe, in *A Tour through the Whole Island of Great Britain*, wrote that until the Great Fire of 1666 the districts of Whitechapel, Shoreditch and Bethnal Green were thinly populated but since then had become 'well inhabited with an infinite number of people'. He estimated the population of these districts at some 200,000, 'where, about 50 years past, there was not a house standing'. He may have exaggerated this figure. The population of the whole London area in 1700 was only about half a million, although it had doubled before the end of the century.

The northern limit of the most densely populated part in the 1720s was St Leonard's Church in Shoreditch, where Fairchild was a parishioner. Just beyond, the landscape was still quite rural: a comedy called *The Walks of Islington and Hoxton*, written and performed in 1657, suggests that these areas were where Londoners went for a breath of fresh air to escape the stresses of the city. West of the main road to Ware and Cambridge was Hoxton Street, leading to Islington and lined with grand houses and estates.

The seventeenth century saw the introduction of the residential square to London – four rows of houses built around an open space. Inigo Jones laid out Covent Garden in 1631, and many of the grand squares of Mayfair, St James's and Bloomsbury date from later in the century. The fashion soon spread north, but on a more modest scale. Two squares were laid out west of Hoxton Street and north of Old Street, running east–west between Shoreditch and Clerkenwell. Hoxton Square was the first, started in 1684 and completed by 1720. Charles Square, to its west, was more ambitious, its houses substantially larger, and the south side remained unbuilt for nearly a hundred years. (One of the original houses still stands today, the headquarters of the London Labour Party.)

London's 'green belt', then no more than three miles from its centre, was a natural place in which to establish market gardens for supplying vegetables to the demanding citizens of the capital. Many of these gardens were close to the Thames in Lambeth, Battersea and Chelsea; but the northern fringes – Hoxton, Hackney and Islington – had their share too. Indeed, in the early sixteenth century, with the demand for fresh produce increasing, so many market gardeners in this area had enclosed parcels of common land for their own use that the citizens of London rebelled. John Stow, in his 1598 *Survey of London*, says that in 1514 a group of Londoners, led by a turner in fool's clothing, marched north out of the city chanting, 'Shovels and spades! Shovels and spades!' Citizens joining the march brought out those implements and used them to dig up the hedges and fill in the ditches that enclosed the market gardens, so that they could once more use the ground for archery or recreational walks, as they had been accustomed.

This direct action had only a temporary effect. When private pleasure gardens became fashionable in the later sixteenth century, it was logical that ambitious landowners should first look to these open spaces to build their versions of paradise. The Hackney estate of the eleventh Baron Zouche became renowned as England's first botanic garden, superintended by Mathias de l'Obel, the French plantsman who gave his name to the lobelia.

Another notable Hackney garden belonged to Sir Thomas Cooke, but J. Gibson, in his 1691 account of several London gardens, wrote that it was 'not so fine at present' and needed money spending on it. 'There are two greenhouses in it but the greens are not extraordinary, for one of the roofs being made a receptacle for water, overcharged with weight, fell down last year upon the greens and made a great destruction among the trees and pots.' (The word 'greens' was sometimes used in that period for plants in general, but more usually it referred to evergreens.) Gibson praised the 'several ponds' in Cooke's garden but added that it was plagued with rabbits.

In his book *Early Nurserymen*, John Harvey records that there had been at least one important garden in Hoxton before the Civil War, belonging to John Noble. For the first nurserymen, the district had the advantage of being close to the City but free from its pollution. It had a plentiful supply of water: the twelfth-century chronicler William Fitzstephen had described its 'fields for pasture and open meadows, very pleasant, into which the river waters do flow', and Stow explained that the water derived 'from diverse springs lying between Hoxton and Islington'. Being on London clay, it was close to several brickworks, a source of flowerpots as well as bricks for the hothouses, or stoves.

The first known nursery in north London was started by George Ricketts soon after the Restoration at the Sign of the Hand, east of Hoxton Street and north of what is now Nuttall Street. His catalogues were packed with trees, shrubs and evergreens, as well as herbaceous plants and flowers. John Rea called him 'the best most faithful florist now about London' and was mightily impressed by his range of 190 tulips. Another seventeenth-century gardening writer, John Worlidge, declared that 'the whole nation is obliged to the industry of the ingenious Mr George Ricketts', with his 'richest and most complete collection of all the great varieties of flower-bearing trees and shrubs in the kingdom'.

Not everyone was so impressed, though. The sceptical Gibson complained about Ricketts' quality and prices. 'He sells his things the dearest and, not taking due care to have his plants prove well, he is supposed to have lost much of his custom.' At the same time he conceded that Ricketts had 'a very good greenhouse, and well fitted with fresh greens', a better stock of Assyrian thyme than any of his competitors and a good supply of the newly fashionable lime trees. George Ricketts was succeeded by his son James, who is known to have supplied plants to the Duke of Bedford at Woburn Abbey in the last quarter of the seventeenth century.

Next to Ricketts' was a smaller nursery owned by Samuel Pearson, who specialised in flowers, although he also stocked tall cypress trees. 'Of anemones he has the best in London,' wrote Gibson, 'and sells them only to gentlemen. He has no greenhouse, yet he has an abundance of myrtles and striped phillyreas [an evergreen shrub], with oranges and other greens, which he keeps safe enough under sheds, sunk a foot within the ground and covered with straw.' Gibson clearly preferred Pearson to Ricketts: 'He is moderate in his prices, and accounted very honest in his dealing, which gets him much chapmanry [trade].' Pearson died in 1701.

The only contemporary source for the precise location of Fairchild's nursery is the naturalist Peter Collinson, writing in 1764 about his London childhood. He recalls that, in about 1712, Fairchild and William Darby, who established his nursery in 1677, 'had their gardens on each side the narrow alley leading to Mr George Whitmore's, at the further end of Hoxton'. Whitmore, a former Lord Mayor of London, had died in 1654 but the magnificent Balmes House, built for him, was still standing at the end of the eighteenth century and was for a while used as an asylum, in which the writer Charles Lamb and his sister Mary had been inmates for a time.

Collinson, recalling Fairchild and Darby, wrote: 'As their gardens were small [Fairchild's was no more than half an acre, according to the rate book] they were the only people for exotics, and had many stoves and greenhouses for many sorts of aloes and succulent plants; with oranges, lemons and other rare plants.' To reach these two temples of exotica, the late seventeenth-century garden enthusiast would have to cross Hoxton Street and walk west down Ivy Lane, so named because it led to Ivy House, a grand mansion that stood at its western end. (Today's Ivy Street is some 40 yards further north.)

The alley to Whitmore's was at right angles to Ivy Lane, running north along the line of what is now Pitfield Street. Darby's nursery was east of the alley. According to Gibson, Darby made his own greenhouses and 'his fritillaria cassa had a large flower on it of the breadth of a half crown, like an embroidered star of several colours'. The writer said he had seen the same flower at the Enfield house of Dr Robert Uvedale, a schoolmaster and keen gardener, but that had been far less impressive. Uvedale was to achieve a place in gardening history as the man who introduced the sweet pea into Britain in 1699, having been sent some seed by a Sicilian monk.

In 1698 Darby was approached to take over as head gardener at the Chelsea Physic Garden but the appointment was not made. He seems to have retired in about 1714, after 37 years in business. In 1716 he married a widow, Martha Winter, and entered into a complex prenuptial agreement that placed most of his assets and securities in a trust, of which Fairchild was one of five beneficiaries, including Martha. After Darby died in 1720 – appointing 'my very good friend Thomas Fairchild' as executor of his will – Fairchild initiated lengthy legal proceedings against Martha and the other three beneficiaries, accusing them of conspiring to conceal the extent of his estate.

After Darby quit the nursery it was in unknown hands for four years, until acquired in 1718 by John Cowell, whose book *The Curious and Profitable Gardener* was published in 1730, a year after Fairchild's own death and Cowell's adventure with his Great American Aloe, described later. The land to the west of the alley was mostly a stretch of open fields that had remained unenclosed since the 'Shovels and spades!' incident of 1514, and was still used for archery practice.

Fairchild's nursery seems from Chassereau's 1745 map to have been carved out of the north-east corner of these fields, across the alley from Darby's. By 1720 another nursery had been established alongside it by Fairchild's friend Benjamin Whitmill, the author of *The Gardener's Universal Calendar* of 1726 – a book that contains one of the most trenchant statements of the sod-turner's creed: 'It certainly redounds more to the honour and satisfaction of a gardener that he is a preserver and pruner of all sorts of fruit trees than it does to the happiness of the greatest general that he has been successful in killing mankind.' Whitmill was clearly a great admirer of his more illustrious neighbour, for when he had a son in 1721 he christened him Fairchild.

Quite late in his career, in about 1725, Thomas Fairchild expanded his operation. He acquired another two acres of land in the southern part of Hoxton, about fifteen minutes' walk from the nursery. The rate book for 1726 shows him

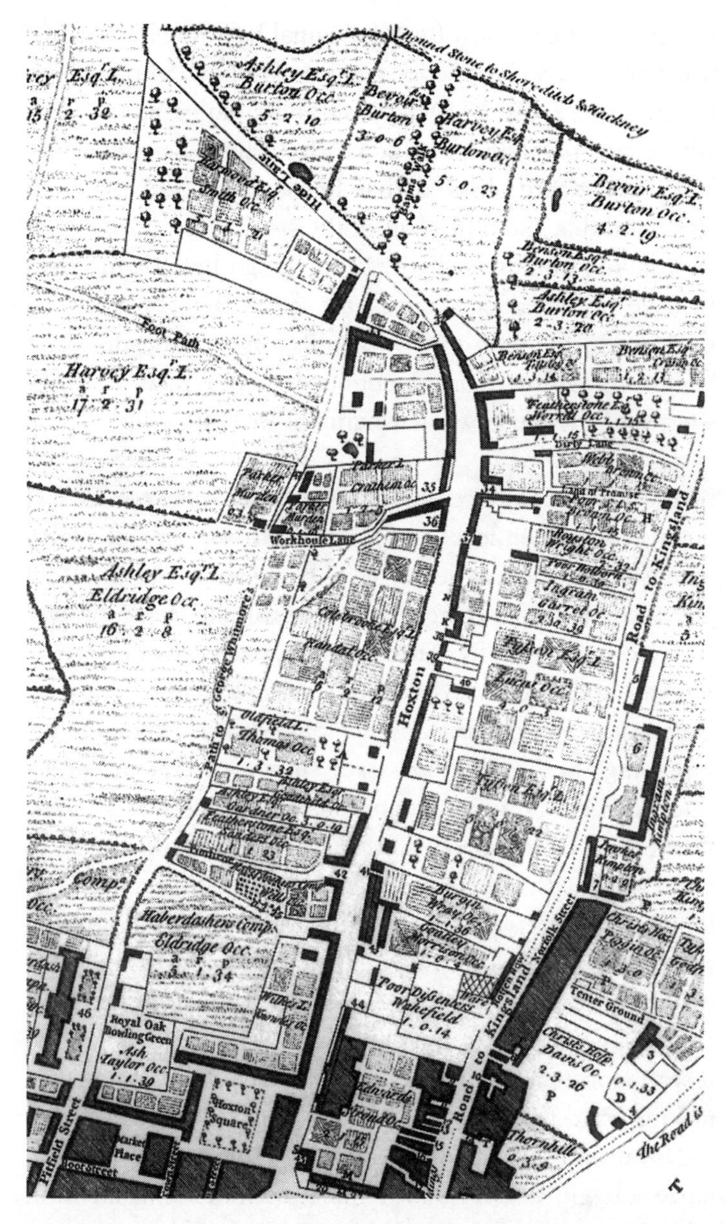

Chassereau's map of 1745. Fairchild's nursery may have been in the area marked 'Parker Murden' (*Guidhall Library, Corporation of London*)

being assessed not just for his original half-acre but for two acres of garden land 'over against the Dog House'. The Dog House, close to Old Street and Charles Square, was where the royal hunting dogs used to be kennelled. By 1727 the record showed that he occupied just one acre there and that the other half of the site had been built over 'by Mr Airs' company'. By 1729 the Dog House land seems to have been reduced to half an acre. After his death his successors Stephen Bacon and John Sampson, and Sampson's widow Anne, continued to pay rates on both sites until the nursery was wound up in 1740.

In the seventeenth century the trade of nurseryman was quite new to London. The first known supplier of plants was William the Gardener, who between 1274 and 1277 provided stock for King Edward I's gardens at the Tower and Westminster. His name suggests that he was primarily a sower and tiller who procured the raw materials for his craft as a sideline, in the absence of any formal structure for supplying the horticultural trade. In the turbulent Middle Ages, gardening was restricted mainly to the monasteries. Owners of large houses were barons and warlords, more concerned with defending their territory than making it look pretty. Flowerbeds would have interfered with the fortifications. What William the Gardener supplied were mainly fruit trees or bushes – pears, peaches, quinces, cherries, gooseberries and grape vines – but there were also roses, peonies and lilies as well as osiers (willow trees) for making baskets.

Not until the sixteenth century did the modern form of nursery begin to develop. Henry Russell's establishment at Westminster dated from about 1530, when grand houses and estates were at last being built to please and impress visitors rather than simply for security. Travellers from Continental Europe would return with tales of the magnificent gardens they had seen, especially in Italy. Not that Russell's stock in trade was substantially different from William the Gardener's 250 years earlier – the same fruit trees and roses, but with the addition of violets, primroses and strawberries, as well as several aromatic herbs for medicine. The range of trees was greater, too, with cypress, juniper, sweet bay and yew now being offered.

Also in Westminster, Banbury's Nursery in Tothill Street opened in 1560 and remained in family ownership for more than a hundred years. The founder, John Banbury, began by growing osiers for the basket trade, but his son Henry broadened the scope of the business and in 1597 was honoured by John Gerard, in his *Herbal*, as an 'excellent grafter and painful [painstaking] planter'. Arnold Banbury, the third generation, was keen on fruit trees and introduced new varieties of damsons, apricots, nectarines, apples and cherries. He had a national reputation: his accounts show him as having sent plants as far afield as Yorkshire and Ireland. The nursery closed when Arnold Banbury died in 1665.

A third Westminster nursery was opened in 1620 by Ralph Tuggie, one of the first of the so-called 'florists' responding to the growing demand for new colours and varieties of the fairly small range of flowers then grown. With the techniques of hybridisation as yet undiscovered, he produced new strains of auriculas, carnations, colchicums and pinks through selection or vegetative reproduction. John Parkinson's *Paradisi in Sole*, published in 1629, mentions his name in connection with several flowers, and there is a drawing of one called Master Tuggie's Princess. After he died in 1633, the nursery was operated by his widow Catherine and then his son Richard, on whose death in 1670 it closed down.

The first documented nursery east of the City was Leonard Gurle's at Whitechapel, on a 12-acre site north of Old Montague Street and east of Brick Lane. Gurle was a fruit specialist, best remembered for the introduction of a late-ripening white-fleshed nectarine called Elruge that is still grown today. The name of the fruit is his surname spelled backwards with an additional 'e'. He acquired the nursery in about 1643, and it continued for more than 30 years after his death in 1685. He was another who combined raising plants with practical gardening and was chief gardener to Charles II from 1677.

In South London, Captain Foster of Lambeth held sway. The hard-to-please Gibson was enthusiastic about his 'many curiosities', his greenhouse 'full of fresh and flourishing plants' and especially 'the finest striped holly hedge that perhaps is in England'. Fairchild, too, was keen on variegated holly.

The Brompton Park Nursery was a mainstay of London gardeners from its establishment in 1681 until 1851, when it was demolished to make way for the Victoria and Albert Museum. From the start, it was a high-powered and ambitious operation, opened by four of the leading plantsmen of the day. Chief among them was Roger Looker, gardener to Queen Catherine (Charles II's wife) at Somerset House. He teamed up with John Field, who looked after the garden at Woburn Abbey for the Duke of Bedford; Moses Cook, responsible for the Earl of Essex's renowned garden at Cassiobury in Hertfordshire; and George London, youngest of the four, gardener to Bishop Compton at Fulham Palace and later to William and Mary, for whom he would redesign the garden at Hampton Court in the Italian style. As well as stocking a broad range of plants and seeds hitherto available only from the Netherlands, the quartet offered a professional garden design service. It was almost certainly the first group practice in the field: all four of them had a hand in laying out the grounds of Longleat in Wiltshire, the seat of Viscount Weymouth.

By 1687, though, Looker and Field were dead and Cook had retired. London persuaded Henry Wise, a young assistant gardener, to buy out Cook's share. Wise had trained in France, where he had absorbed the classic design principles of André Le Nôtre, creator of the gardens at Versailles, Chantilly and Vaux-le-Vicomte. Under London and Wise the nursery went from strength to strength, although to judge from Gibson's description it did not have much of a garden, being 'chiefly a nursery for all sorts of plants, of which they are very full'. It had a large and impressive greenhouse, where the tender plants from William III's Kensington Palace were overwintered. London lost his post as royal gardener when Queen Anne came to the throne in 1702, but Wise was appointed to replace him, so the firm's prestige and patronage did not suffer.

London was now free to tour some of the grandest estates in the land, including Windsor, Chatsworth and, most notably, Blenheim, offering advice on design and suitable plant material. Naturally, the business went to the Brompton Park Nursery. Stephen Switzer, who had trained at Brompton Park, sang his praises, though with some qualification. Fruits were 'his masterpiece', and for other growing things 'he certainly had as much knowledge as any one man living'. Switzer considered garden design to be London's weakest point:

> Though he might not always come up to the highest pitch of design, yet that might be attributed to the haste he was generally in; and it can be no great

> blemish to his character that he was not the greatest person in everything, when 'tis surprising to find he could possibly know so much.

The essayist Joseph Addison would brook no such criticism. He waxed lyrical about the pair in the *Spectator* in 1712:

> I think there are as many kinds of gardening as of poetry: your makers of parterres and flower gardens are epigrammatists and sonneteers in this art; contrivers of bowers and grottoes, treillages [trellises] and cascades are romance writers. Wise and London are our heroic poets.

Half a century later many of George London's grandiose designs would be buried beneath the 'natural landscapes' of Lancelot ('Capability') Brown and his followers, who revolutionised our idea of what a country estate should look like. But while elaborate formal beds were still in style, the demand for plants, flowers, shrubs and straight avenues of trees was phenomenal. London and Wise, having had a hand in creating the demand, were only too happy to supply the raw materials, and as a result they prospered. In 1715 Switzer estimated the value of the plants at Brompton Park Nursery at between £30,000 and £40,000, 'perhaps as much as all the nurseries of France put together'. When Wise died in 1738 his estate was worth over £100,000. Translating that into today's terms by using the conventional multiplier of 100, it would come to £10 million.

In the emerging world of nurseries, the hundred-acre Brompton Park decidedly represented the big time, the closest equivalent to a modern, full-service garden centre that the late seventeenth century could offer. In 1705 it was estimated to contain ten million plants. When Fairchild took his far-reaching decision to go into the same business, he knew he would be up against a strong, perhaps an overwhelming competitor, as well as the 15 or so other nurseries then established in the London area. What did he have to offer that could begin to compete with such a smoothly run, well-connected and profitable business?

He was a down-to-earth man, not easily daunted. The market was expanding rapidly and he thought he detected a gap in it. Replicating fancy Continental parterres in large country estates was all very well, but surely there was also room for a business that put more emphasis on the plants, especially on developing new forms of them. London and Wise were artists, for sure, and up to date with all the latest design trends; but were they scientists? Did they really understand why flowers look the way they do? Fairchild was not a scientist in the strict academic sense, for he had left school to take up his clothworker's apprenticeship. But he was practical and, above all, he had an intense curiosity about the workings of nature.

He would never, from his modest half-acre, be able to provide ornamental plants, such as box, in sufficient quantities to fill the requirements of the large estates like Chatsworth and Blenheim. So he would have to specialise in rare and exotic plants, some from overseas and several difficult to rear, allowing him to charge a high price to make up for the low volume. To do that, he needed to call on his skills of observation and careful record-keeping – noting which soils suited which plants, how much heat they needed to prosper and how they could be persuaded to reproduce themselves. Through cultivating that expertise, and through his unfailing curiosity, he was destined to leave an important and permanent legacy to gardeners of the future, a legacy that would revolutionise the practice of horticulture worldwide.

CHAPTER THREE

A Little Knowledge

Notwithstanding I have been about forty years in the business of gardening, I find the art so mysterious that the whole life of a man may be employed in it without gaining a true knowledge of everything necessary to be done.

THOMAS FAIRCHILD, *THE CITY GARDENER*, 1722

References to Fairchild and his nursery by contemporary writers tell us that he specialised in trees and vines, which still accounted for a significant proportion of the garden market. In 1724 *Parker's London News* reported that he 'has been many years collecting foreign grapes and other rarities in his way'. It singled out 'amongst his curious collection a bunch of grapes half a yard in length and above an ell [45 inches] in circumference, and the grape in proportion as large, which far exceeds everything of the like nature in England'.

Bradley, in his *Philosophical Account of the Works of Nature* (1721), wrote that Fairchild's nursery held 'the greatest collection of fruits that I have yet seen', adding that they were 'so regularly disposed, both for order in time of ripening and good pruning of the several kinds, that I do not know any person in Europe to excel him in that particular; and in other things he is no less happy in his choice of such curiosities, as a good judgment and universal correspondence can procure'. Bradley enthused still further in his *General Treatise of Husbandry and Gardening*, a monthly journal that he published in 1722 and 1723: 'For eating grapes, I have hardly tasted better in any part of

Europe where I have been than of these sorts at Mr Fairchild's garden, which had only the benefits of common walls to ripen them.'

Much of what we know about Fairchild's nursery comes from Bradley's writings. It is from him, for instance, that we learn that the ground was heavy London clay, in flat countryside. Bradley was an intriguing and controversial character, but little is known of his origins. He was probably a few years younger than Fairchild, and is thought to have been only 26 when elected to the Royal Society in 1712. Energetic and entrepreneurial, he was one of the first botanists to make a thorough study of succulents. Between 1712 and his death 20 years later he wrote 24 books, mainly on horticulture, and between 1721 and 1724 produced his journal, originally called *The Monthly Register of Experiments and Observations in Husbandry and Gardening*.

He has been described as Britain's first gardening journalist. This is not necessarily a compliment, for like many who follow that trade he was regarded with suspicion in respectable circles – and justifiably, for although a skilled and enthusiastic botanist and gardener, he seems also to have been an out-and-out charlatan. Some twentieth-century scholars have sought to defend him, notably Dr H. Hamshaw Thomas in a paper to the British Society for the History of Science in 1951. Declaring that Bradley was 'greatly in advance of his time in his outlook on plant life', Thomas appeared to believe that a man with such a creditable thirst for knowledge must have had some redeeming features; but the record suggests otherwise. Either he was desperately imprudent and unlucky with money or he was a complete rogue.

Bradley's finances were permanently precarious. The best documented year of his life is 1714, when he went to the

Netherlands and exchanged letters with his friend James Petiver, an apothecary and botanist who was also a friend of Sir Hans Sloane. Petiver had made a similar trip three years earlier and gave Bradley the name of several Dutch contacts, some of whom would not have thanked him for the favour. For Bradley's thoroughly entertaining letters, while they portray their author as a congenial companion for an evening of yarn-spinning in a tavern, show him distinctly lacking in prudence or probity; a congenital conspirator and fraudster with chronic financial problems that he would try any ruse, however dishonourable, to resolve.

The Dutch project even began furtively, with Bradley inviting Petiver to meet him at the Three Fighting Cocks, a tavern in St George's Fields in Southwark, an area with a dubious reputation on the other side of the river from their usual haunts. He instructed the apothecary to say nothing openly there about his projected trip. A later letter describes how he went under the alias George Grant, disguised in a black wig, and had arrived 'clear of all danger'. An American scholar, Frank Egerton, has suggested that all this was a product of Bradley's psychological need to act in a clandestine manner. A more likely explanation is that he was escaping his creditors.

His purpose in the Netherlands was partly to collect plants for Fairchild, mainly from the renowned botanic garden at Leyden, which contained many items newly introduced into Europe, derived from the expeditions of Dutch explorers and colonists in the West Indies and Africa. In return, Bradley would supply Dutch collectors with a few of Fairchild's specialities, such as variegated ivy and sedum. He was also collecting 'curiosities', including dried insects and pressed flowers, for Petiver, Sloane and others, and doing some botanical paintings that from time to time he tried to sell.

Petiver was an especially avid collector of curiosities. Sloane wrote that he had acquired a greater quantity than any man before him – but did not look after them too well: 'He did not take equal care to keep them, but put them into heaps, with sometimes small labels of paper, where they were many of them injured by dust, insects, rain, etc.' Beginning in 1696, Petiver began publishing lists of his curiosities, a hundred at a time, in the Royal Society's *Philosophical Transactions*, and many can be inspected today at the Natural History Museum in London.

Some 300 years on, it is easy to dismiss these eighteenth-century doctors and gentlemen as dilettantes motivated only by the desire to accumulate odd objects to impress their friends. To picture them gathering and dispatching dead beetles, dried flowers and such, and submitting them to solemn analysis, seems more than a little quaint. In fact it was only by collecting and examining as many examples as possible of plant and animal life that their common characteristics, their means of reproduction and their relationships to each other could begin to be determined, and the foundations laid for later scientific discovery. If Fairchild, like his contemporaries, had not familiarised himself with countless living and preserved flowers, he would not have had the basic knowledge required to carry out his experiments in hybridising and pruning.

Bradley's letters from the Netherlands go into some detail about his progress in these several endeavours, the practical shipping arrangements (in one he chastises Fairchild for being half an hour late at the London Custom House to pick up the consignment) and his flattering confidence that Fairchild will give all his samples the tender care they deserve. All these matters, though, are coincidental to the theme that runs right through the correspondence: his chronic lack of money. 'I spend all my time in collecting of plants from the garden and it would be a great pleasure if I could get anything by it,' Bradley told Petiver grumpily.

That letter, dated 4 July 1714, went on to describe his most extraordinary money-making ruse: to pretend to be a doctor. Given the state of medical knowledge at that period, and his growing botanical expertise, he was probably able to make almost as good a fist of it as anyone, but he lacked any formal qualifications. As he tells it, his first venture into medicine came through a misunderstanding on the part of the locals – although one that, given his parlous financial plight, he naturally did not try over-zealously to correct. The disease that he was first called on to cure confirms that the reputation of Amsterdam has remained largely unchanged over the centuries:

> The people here will have me a doctor whether I will or no, and to carry on the jest I desire you will favour me with the recipes and medicine for the venereal distemper, for here are some English gentlemen who now and then receive that favour and won't be persuaded to take Dutch pills. I know a little of the method where it is but . . . I want the recipe of camphor pills and white mercury at the first, the decoction [concentrate] of woods.

Petiver must have responded positively to Bradley's unorthodox request, for a few weeks later Bradley was writing cheerfully that his deception, posing as 'the doctor in the black wig', was going well. One of his 'patients' was about to visit London, and Bradley had given him a letter of introduction to Petiver. 'Pray carry on the jest,' he urged Petiver, 'for 'tis like to prove a good one, for Dr Ruysch teaches me anatomy and with my own smatch of botany I begin to smell a little quackish.' The cover story to be sustained by his English friends was that he was 'a doctor in physic, bred at Oxford, in a black periwig, a great virtuoso, and now travelling for the improvement of natural knowledge'.

Fairchild, who probably knew little of these larks, was still, in his methodical way, packing plants for Bradley to pass on to friends and clients in the Netherlands and doggedly sorting through the mixture of rare and familiar specimens that Bradley was sending him in return. Bradley wrote to Petiver:

> For Fairchild's care I have sent 13 sorts of aloes and several other varieties of that kind which are all new to England, and in a little phial three coffee berries which he must open with care and immediately put the seeds they contain . . . separately in the ground about an inch deep, plunge the pots into a hot bed and given them a little water. They will not come up if they are kept till next year.

Bradley was intrigued by coffee and was later to write a paper about its cultivation in Britain, although this practice never proved commercially viable. As well as the three seeds, he hoped to send later a plant two or three years old 'but money must come first or my harvest will be spoiled'. The money must have arrived, for a week later, on 31 July, Bradley wrote to say that the coffee plant had been dispatched.

That letter was written from The Hague, where he had

been taken by another of his patients, who 'trials me nobly, will carry me back and I believe pay me well'. He thanked Petiver for his earlier advice and was keen to advance his medical skills still further:

> I received yours with the insects and my doctor's degree. I have two patients and take upon me at a huge rate. I have learned to write the recipes [for remedies] by heart but you have forgot that for the drink made of sassafras, guiaicum [a West Indian tree] and sarsaparilla, which I hope will come in your next, with some recipes for a fever or other common distemper. Likewise how to take away a swelling in the testicles and a cataplasm [poultice] to raise bubos [swellings of the glands – a common symptom of the plague] when they begin to appear, with what caustic to apply when it is come to a head, etc.

Not every aspect of Bradley's trip was going as well as his medical practice. Always keen to forge new connections with the aristocracy, in The Hague he had sought to call on the third Earl of Burlington, who had stopped there on his way to study Palladian architecture in Italy. 'Fortune still frowns,' he reported glumly. 'His Lordship went out as I came in, purposely as you may guess to disappoint me. He does not return till tomorrow.' Yet he remained optimistic: 'I have met many English gentlemen who are posting for Italy and it may be that I shall wriggle myself into some of their favours and take the trip with them, but that much depends on a healing letter from your side of the herring pond.' Sadly, he did not wriggle convincingly enough, and Burlington – no doubt prudently – left without seeing him.

Petiver's actual replies to Bradley's letters have not

survived, but the drafts of some of them appear in one of the apothecary's notebooks. In the longest, dated 3 August 1714, he reported 'the sudden and surprising death of our most gracious Queen [Anne] who expired on Sunday morning', marking the start of the Georgian era. He reports with some relief that the Elector of Brunswick (George I) had been proclaimed king by agreement of the Lords and Commons, which had been in session ever since the queen's death. Petiver was clearly as anxious as most of London's great and good that the crown should remain in Protestant hands and not revert to the theologically suspect Stuarts.

Much of the letter was taken up with discussing Bradley's botanical deliveries. Fairchild had received the live plants and seeds safely, but Petiver was disconcerted by the condition of the dried specimens, which came somewhat damaged and without labels, 'so they are of little use to me and I shall therefore reserve them till you return'. Several he recognised as being common English wild plants; others were the same as those being grown at the Chelsea Physic Garden and other places in London, 'which you need not have given yourself the trouble to collect there'. He reminded Bradley that this was the season to garner flower seeds, 'which may be easily done by carrying with you small paper bags whenever you go into the garden to put them in'. He added: 'I hope you may send [them] to Fairchild to raise for you and bestow on other friends to their advantage.'

Then came the medical advice that Bradley – and presumably his patients – had been awaiting so anxiously: 'When you meet with a swelling in the testicles you must purge and bleed, largely anointing the part with oil of sambuci, i.e. oil of elder, and when bubos begin to suppurate . . . soap lees boiled to a consistency.' This should take effect in four or five hours. Other remedies included 'powders of crabs' eyes, coral or oyster shells, which will do as well to be taken in a cordial'.

Despite his new source of income, Bradley's money problems did not go away. Petiver, and perhaps Fairchild, appears to have sent some funds, but not enough to keep his head above water were it not for the assistance of Pancrass, the local burgomaster. 'Many aloes cost 100 guilders,' he complained, 'even the cuttings I have sent and which I could not compass without the friendship of the Burgomaster. Whoever has any of them must pay and at the same time give me the assurance of the first offset . . . Pray send me what I writ for last post and forward the guelt, otherwise I may prepare for a lodging in the Stadhouse [prison].'

Reporting the arrival of some plants from London after a long delay ('a miracle', he enthused) he disclosed that he had difficulty in getting them unloaded because he did not have enough money to buy the captain a drink. The day was saved again by his good friend Pancrass who, ordering that the goods be seized from the ship, hauled the captain up before the magistrates. At the end of the day Pancrass consoled Bradley for the trouble by treating him to 'a melon and a very good burgundy' in his chambers.

Bradley was impressed with the quality of the plants that Fairchild shipped, contrasting them with those from other nurseries, many of which were dead or 'spoiled' on arrival. So delighted were the Dutch gardeners to receive them that they treated him to another 'very good burgundy'. But still the financial clouds refused to lift. 'In a word,' he wrote, 'if money does not come I must turn another Ovid and metamorphose coats and waistcoats into guldens, or else with Icarus's wings fly 'til I am melted body and boots.'

Small wonder, then, that he was indefatigable in his attempt to drum up new business, sending an instruction to his wife, through Petiver, to 'acquaint the Duchess of Beaufort I am surrounded by curious plants and only wait her Grace's commission'. He appears to have been successful in that, at least, for a later letter discusses arrangements for shipping plants and curiosities to the duchess's Chelsea house – she seemed especially fond of aloes – although he apparently had difficulty in getting her to pay for them: 'I desire you will mention to the Duchess the expense I have been at, to let her know how dear plants are on this side.'

Came August and Bradley, still in Amsterdam, was observing with some indignation that 'the Hollanders are not much affected at the death of our Queen'. He penned a travelogue on the Netherlands, in terms any Londoner would understand: 'The Hague is St James's or the court end of the town. Leyden is the Temple, Rotterdam the City. Amsterdam is Wapping [the dock area] and Utrecht is Westminster, where the capital affairs are transacted. In my late voyage I was much diverted with a true original Dutchman. He is about seven feet high and as fat as a hog.' He also included a pretty painting of a green butterfly, one of several given to him by a ship's doctor who had acquired it in Batavia (Indonesia).

After a whole summer of such letters, Petiver must have known that, after the pleasantries, Bradley would inevitably get down to the business of money. He was pleased to hear that the last batch of plants had arrived safely: 'I don't question the novelty of them in England and the transports of Mr Fairchild at the sight of so many strangers.' But he was more concerned with extracting payment from the

Duchess of Beaufort: 'I desire that the coffee tree may be sent to the Duchess and a copy of the bill enclosed . . . I have had credit with an English gentleman who has lent me almost 200 guilders to pay my landlord and the Customs and other charges, but I have been so long without return I can ask him for no more, and unless they come speedy your humble servant, the doctor, must make use of hocus pocus.'

To this end he had put up a notice at the end of his street: 'Mr Bradley, surgeon . . . may be advised with every day from 8 in the morning till 2 in the afternoon at the first house on the left . . . and at other hours at his home.' But not all his remedies were going well:

> I have had a patient who about ten months ago got a mishap. I followed the prescribed method but the camphor pills, only three in a day, made him piss almost every minute, and with the greatest pain, so that he was quite brokenhearted, and that which was worse, as soon as he had got a little relief from them a violent colic seized him . . . It would make you laugh if you had seen my grave way of managing him and how much the good opinion of the physician works the cure.

So he was perfecting his bedside manner and, when that and science failed, he turned to faith-healing:

> I on my side, when I don't know what to do, say I am not for giving more physic than is absolutely necessary, and as I find nature has yet strength enough to help itself I recommend a little patience and all will be well. To make up for all, I attend six times a day and

> receive the thanks of the whole house for my great care and wise conduct.

The following month he needed the attentions of a physician himself, telling Petiver that he was suffering from the ague, or fever: 'I have advice enough from the best doctors to eat sugar till my guts crack.' In his enforced idleness he was engaged on a new enterprise connected with another scientific controversy of the time: the difference between the Julian calendar used in Britain and the Gregorian calendar that had been taken up by the rest of Europe in 1582. 'I have composed a calendar of 13 months in the year . . . which reconciles the difference between the two styles.' The extra month would be 'addressed to our King as July and August were to emperors'. (It was not until 1752 that the situation would be resolved by cancelling 11 days in September.) He promised to send a copy of the proposed new calendar to Petiver with the next batch of botanical specimens, including 'a handsome wild plant of the striped bay for Mr Fairchild'.

Naturally, he was still short of funds: 'As soon as I can get money I will leave this cold country for Paris and there study medicine.' But by October he had at least recovered from the ague: 'I cured myself with a drink I invented: a quart of boiled water with cinnamon, a quart of white wine and two lemons. I drank no other drink and that always hot. I have cured my landlord and his wife with the same. So much for my abilities and inabilities.'

Before the end of the year, on his way home, he did go briefly to France, where he suffered a misadventure that illustrates the hazards of transporting live plants at that time. In France he bought several hundred vines, but because of delays in shipping and then at the Custom

House only one in eighty survived. He drew an apocalyptic moral a few years later in his *General Treatise of Husbandry and Gardening*:

> Unless there can be some way found out which may allow free passage for things of this nature which cannot bear delays, I fear we shall make few additions to our plants in England, whether useful or curious. Nor was this the only time I have suffer'd at this rate, for some time before the state of Amsterdam presented me with above 150 different sorts of curious and valuable plants, which were strangers then in England, i.e. I had not seen them in any of our English gardens of note. Mr Fairchild of Hoxton was my correspondent but there was so much difficulty to get them landed that above two-thirds of them were destroyed.

Among the problems of sending plants by sea was that they usually had to be kept on deck, and thus impeded seamen as they moved around at speed to adjust the rigging and to secure the ship when it arrived in port. Plants were therefore unpopular with captains, who would usually prefer almost any other freight. So the plants would often languish for weeks on the quayside in their pots and containers, and only the toughest would survive.

However, Bradley's sojourn overseas had at least succeeded in broadening his knowledge of the range of flowers and fruits that it was possible to cultivate. A few years later, in a letter to a friend, he recalled one discovery that perhaps was worth more than the postscript he gave it: 'There is a sort of pea in Holland which has no skin within the shell, so that the people eat them shells and all, as we do kidney-beans [French beans]; 'tis very sweet and very profitable and I hope to introduce it.' That may have presaged the arrival in England of that adornment of today's fashionable dinner table, the sugar pea, or *mangetout*.

In 1717, back in London, only a friend's generosity saved Bradley from debtors' prison. He had been working for James Brydges, the first Duke of Chandos, superintending as many as 40 garden labourers at his splendid house at Canons, north of London near Edgware. For a time Canons was the cultural and horticultural pinnacle of polite society. George Frederick Handel was its resident musician and composed 12 anthems for the duke. In 1722 John Macky, in *A Journey Through England*, described the magnificence of the gardens: a parterre with gilded vases and many statues; see-through iron balustrades dividing the sections instead of the conventional brick and, in the vegetable garden, 'bee-hives of glass, very curious', which I suppose may have been see-through hives but were more probably early bell cloches. Garden records speak of acorns imported from Italy, flower seeds from Syria, fruit trees and exotic birds from the West Indies.

Sadly, by the time of Macky's journey, Bradley's connection with Canons had been severed. According to the duke's journal he was dismissed because he had been 'mismanaging the hot-house, the physic garden and the sums entrusted to him', which would certainly have been in character. Worse was to come for the would-be garden manager to the rich and famous. In a plaintive letter to Sir Hans Sloane on 23 June 1722 he referred to 'the unfortunate affair at Kensington whereby I lost all my substance, my expectations and my friends'. The details of this crisis remain a mystery, but Dr Hamshaw Thomas suggested that he might have been ruined, like many others, by the frantic speculation surrounding the South Sea Bubble, which burst in 1720. Certainly, it was the kind of get-rich-quick scheme that would have appealed to him. Kensington could have meant Kensington Palace, then the residence of George I, some of whose courtiers were implicated in the Bubble scandal: Bradley claimed more than once to have 'friends at court'.

A more likely explanation lies in his friendship with Robert Balle, a merchant with a large house on Campden Hill in Kensington, where Bradley was apparently given the run of the garden after his dismissal from Canons. Balle was the man who proposed him for fellowship of the Royal Society, and Bradley dedicated a book to him, but their relationship seems to have been curtailed by the 'unfortunate affair', and it may be significant that Balle moved out of the Campden Hill house at about this time.

Whatever the unhappy event in Kensington was, Bradley's publishing ventures did not seem to have been doing as well as he had hoped. In his letter to Sloane, he wrote that since the crisis 'I have endeavoured to support myself at the public expense, and the public have been so good natured to give me sufficient support to enable me to pay about £200 debt besides what I have lost by booksellers, which would have done as much more.' Despite these losses, Bradley said he would continue to publish his monthly journal 'for a little while', to give him a chance to respond to queries from readers of earlier volumes. He was also planning a book about succulent plants and had already obtained some engravings to use as illustrations.

Although bookshops abounded in London in the early eighteenth century – mostly in the area of St Paul's Cathedral – publishing was a hazardous business. As

Michael Treadwell pointed out in a paper in the *Library* in June 1982, publishers invariably doubled as booksellers, supplying stock to other, non-publishing bookshops as well as to individual customers. It was often up to the author to raise the money for publication, either from his own resources, by selling 'subscriptions' (copies in advance) or through sponsors who, in return for their generosity, would have the book or pamphlet dedicated to them and would usually come in for some extravagant praise into the bargain. Thus, the publisher seldom took much financial risk – a point noted sourly by many writers, including one in the *Gentleman's and Lady's Palladium* of 1752, who described a publisher as 'a sort of bookseller in miniature, but guilty of far greater extortion since he neither advances any money nor runs the least hazard, and yet is hardly satisfied with 30 per cent per month for vending another's property' – a complaint of authors through the ages.

Most of Bradley's early books and pamphlets were published by W. Mears, a bookseller at 'the Lamb, without Temple Bar', whose best-known author was Daniel Defoe. In 1721 Mears published, alone or in partnership with others, four of Bradley's works, but a fifth, *A General Treatise of Husbandry and Gardening*, went to a new publisher, John Peele, of Locke's Head in Paternoster Row, suggesting that Mears may have been the bookseller whom Bradley referred to disparagingly in his letter to Sloane. But if there was a dispute it was not terminal, because Mears began to publish Bradley again in 1729. In the intervening years Bradley used a variety of publishers, among them Thomas Woodward, who was described as the publisher to the Royal Society. Woodward and Peele were the publishers of Fairchild's *City Gardener* and were also involved in producing a newspaper, the *London Journal*.

In his 1722 letter to Sloane, Bradley confided that the Kensington business had been such a blow that he was considering leaving the country: 'I have some friends at court who do not care I should go abroad, though my inclinations are for it, even into the most dangerous country. But to live upon expectations at home is as bad as it can be to venture one's life among savages abroad.' To avoid both eventualities, he sought to establish 'a garden of experiments for general use, such as I should have accomplished if I had not had the Kensington misfortune, and by that means I might gain an improving settlement and I hope to do my country some service without restraint of booksellers'.

Although the letter ends with a request for Sloane's advice, it seems a thinly disguised plea for funds. These were apparently not forthcoming, so Bradley embarked on a different course. Instead of risking his life among savages abroad, he decided to brave the shark-infested waters of English academic life. He repaired to Oxford and sought to be appointed Professor of Botany – a post that carried with it no stipend but a great deal of prestige as well as the stewardship of the oldest botanic garden in the country. On 17 October 1723 he wrote to Sloane asking him to intercede on his behalf. The bid failed, but in 1724 he was chosen as Professor of Botany at Cambridge, a post he kept until his death.

His detractors alleged that he won the appointment on false pretences, having given assurances to the university that he would finance through his 'private purse' a university botanic garden to rival Oxford's. His supporters say the story was put about by disappointed rival candidates for the post. Whatever the truth of the matter, it is apparent that his private purse contained only loose change, and the establishment of the botanic garden had to wait until after his death. Yet there is no doubt that he genuinely wanted it, for in 1725 he wrote a detailed – if optimistic – description of what it should contain. He saw it chiefly as a place where experiments could be carried out to allow farmers and gardeners to see the results of new techniques, and to compare different ways of raising crops:

> Besides collecting such plants as are used in physic, and choice vegetables from foreign countries, a little room may be spared for experiments tending to the improvement of land, which may be the means of increasing the estate of every man in England.

He also had an international vision for the garden as a place where ways could be tested of growing commercial crops of economic benefit to tropical countries. He quoted the case of coffee, originally found in Batavia and then brought to Amsterdam, 'where, after little time, they raised several hundreds and sent them to Surinam and Curaçao, in the West Indies, from whence, I am told, they receive a good freight of coffee every year'. He thought the lesson could be applied to Britain's plantations in the Americas: 'I can see no reason but that we may render them more advantageous than they are at present, by sending to them many plants of use, which will grow freely there and may be collected and prepared for them in such a garden as I speak of.' As S. M. Walters points out in *The Shaping of Cambridge Botany*: 'A century and a half later this was precisely the role played by the Royal Botanic Gardens at Kew in relation to tropical and subtropical crops in the British Empire.'

The Cambridge appointment clearly did not solve Bradley's financial problems, for in November 1726 he was writing to Sloane again, offering him 'a saffron kiln of the best sort' for eight shillings. He had bought it on his way from Cambridge to London. He was still having trouble with his publishers and yearned for a wealthy patron:

> I want very much to get out of the booksellers' hands and not to write so much as I do for the public, but I cannot live without making my remarks and observations in natural history and drawing and painting whatever is curious and comes my way. I would much rather have what I do in the hands of a gentleman than to be published. If I could find anyone who would give as much as a bookseller I would sell them curiosities enough.

Other letters from Bradley to Sloane, which have survived with Sloane's papers, invariably beg for his support in financial or other ways – such as the loan of books to help him with his botanical writings. On 29 September 1727 Bradley wrote acknowledging Sloane's help in paying the stamp duty on his journal. He said he was seeking more booksellers to handle it, so that the income could clear his debts. He vouchsafed that £100 would do the trick, so that 'I should then appear like other men', and added in passing that he was still looking for 'some great personage', to underwrite his proposed botanic garden in Cambridge.

Sloane stubbornly refused to take the hint and in 1727 appears to have lost patience with Bradley, apparently accusing him of not supporting his candidature for the Presidency of the Royal Society, or his slate of candidates for the committee, following the death of Sir Isaac Newton. On 1 January 1728 Bradley wrote him a snivelling letter denying the assertion and blaming ill-willed mutual acquaintances for the slander. 'I now fear some malicious wretch envied me your favour and has endeavoured by this means to break the pleasure and advantage I had in it.'

He went on to speculate on the wretch's motive:

> I am envied by some people who would not have a physic garden erected at Cambridge and are jealous you would contribute towards it; and for that reason would even tell you mischief rather than truth. It would be generous to acquaint me who is the robber of my reputation on this occasion; or that you would look upon him if he cannot clear himself from my charge against him, as a man that would pick your pocket.

Whether Sloane named names is not recorded, but relations between the two men were repaired well enough by September 1729 for Bradley to ask Sloane once more for emergency funding. Again he could not raise the money for the stamp duty on his journal, and without it he would have to abandon the issue due to go to press the following week. 'Should I miss the day of publication the whole design will be destroyed. I have not friends in town at present to lay down money enough and beg you will favour me with your assistance . . . Those who are to be sharers in my paper in the future do not meet till the next week, but all my hopes of having a settled income from them will be destroyed if this should miss the publication.' At the foot of the letter Sloane wrote: 'Sent him a guinea.'

Bradley was clearly willing to try anything to garner a

crust. In the edition of the *General Treatise* for August/September 1724 he seems to have been advertising his services as a garden designer, using Fairchild's nursery as a base. He complained that most designers did not have the ability to convey their intentions accurately through drawings: 'It is true that by shading of a draught one may in some sort represent hollows, slopes, terraces, etc., so that the workmen may understand how to work from it; yet the gentleman for whom it is made can never rightly frame an idea from such a draught of what it will be, and how it will appear when it comes to be finished.' To solve that, he suggested a model be made, so that 'we may observe the risings and sinkings of the ground, the terraces, the hedges and every other part as it will appear to the eye . . . so whatever offends the eye in a model must necessarily offend in the work itself'.

He worked up more and more enthusiasm for the idea:

> Considering how cheap a model might be made of a garden, and how much money it might save a gentleman in alterations . . . I wonder nobody has yet had models of gardens made. If it is because it has not yet been thought on, or because it is not known where such things can be made, I shall inform my reader that I have instructed one in the method of making them, and embellishing them in a proper manner, who may be heard of at Mr Fairchild's at Hoxton.

Around 1730, in an apparent final attempt to solve his money problems, Bradley married a widow named Mary, but soon ran through her fortune, as was explained in her petition to Sloane after her husband's death in 1732: 'Soon after the marriage [she] was obliged to sell and dispose of her household goods and plate for payment of his debts and other encumbrances and was since obliged to live with him in ready furnished lodgings in and about London.' During Bradley's final sickness, Mary had to sell her clothes to support him. 'At the time of his death she had nothing left and has a child to maintain.'

Until he died, Bradley continued his attempts to 'raise the wind', as he put it, from patrons to fund the Cambridge garden. It did not help that he was unpopular with the university establishment: he spoke no Latin or Greek, the principal languages of scholarship, and was criticised for scarcely ever delivering a lecture. This is odd, because the texts of several lectures have survived, although that does not necessarily mean that they were all delivered. As Walters surmises, it may have been that the content of the lectures was considered too practical and insufficiently scholarly for an audience at a university where botany had until not long previously been regarded as essentially a branch of classical studies. After all, his critics reasoned, the students were not bent on becoming gardeners, so they were not too interested in how to prepare ground for planting.

Putting aside his obvious weaknesses, a strong case can be made that Bradley was a pioneer plantsman and botanist, combining an enquiring mind with a considerable amount of gardening know-how – but even here it is hard to take at face value some of his claims about his achievements. For instance, in April 1724 the *Weekly Journal* reported that 'by his extraordinary skill in gardening' he had managed to grow cherries that ripened at the beginning of April, and that other fruit and produce – peaches, nectarines, plums, apricots, figs, grapes, pears and beans – were almost fit to gather at that time also. The fruit was 'as large and as well flavoured as if they had been produced in a natural way', and to prove it he had presented the Princess of Wales with a parcel of French beans, roses and other flowers about three weeks earlier – which would be in mid-March.

Bradley was clearly a man of some ingenuity and skill but, never having found a way of making his talents pay, lived permanently on the brink of disaster. Sadly, no direct correspondence between him and Fairchild survives, but as this book progresses we shall gather enough from Bradley's publications, and references in letters from acquaintances, to grasp something of the nature of the friendship that existed between them for 20 years or more.

CHAPTER FOUR

Fairchild's London

My design is only to instruct the inhabitants of the City, how they may in little arrive at the knowledge of managing and delighting in those gardens which their present industry leads them to retire to, when their business has given them sufficient fortunes to leave off trade; and I doubt not but, from my experience, I may add some benefit to those who have already begun to show their love for gardening, even in the smallest way, let it be never so little.

THOMAS FAIRCHILD, *THE CITY GARDENER*, 1722

London's professional gardeners had been organised loosely since at least the mid-fourteenth century, when they petitioned the Lord Mayor to be allowed to continue to sell their produce outside the east gate of St Paul's churchyard, as had been their long-established custom. The request was refused, and they were removed instead to a spot south of the cathedral – the first of several slights by the City authorities directed against the gardeners, who never enjoyed the status of the great medieval livery companies, or guilds.

The Worshipful Company of Gardeners had to wait to receive its first royal charter until 1605, by which time the art of decorative gardening was beginning to rival in importance the production of fruit and vegetables. Its coat of arms portrayed a man clothed only in an animal skin, digging barefoot with a spade. Two female figures bearing cornucopias of fruit supported him on either side, and a basket of flowers and fruit topped the design. Beneath this was the motto 'In the sweat of thy brows shalt thow eate thy bread.'

A second charter, of 1616, forbade anyone who was not a member from gardening commercially or selling home-grown produce within a six-mile radius of the City. However, the company still experienced difficulty in getting full recognition from the authorities at a time when the other craft guilds were vying for precedence and privileges. Not until 1659 was it formally recognised by the City Corporation and its members allowed the freedom of the City. Even then it was discriminated against by not being granted its own livery, as other guilds had. This was a blow to the gardeners' esteem and influence, because without a livery the company could not appoint liverymen, and only liverymen were allowed to vote in elections for London's mayor and sheriff.

By 1649 the company controlled large market gardens just outside the City, employing some 1,500 labourers and 400 apprentices. Towards the end of the century, as the size and number of gardens increased, hundreds of unqualified labourers came to London from other parts of the kingdom to work in the gardens of large houses. This worried the Gardeners' Company, who feared that their properly trained members would be swamped by this influx of the unskilled. A member of the company published a letter complaining about 'higglers [itinerant tradespeople] in plants, who impose upon the buyers rotten or decayed plants, trees and ferns without any possibility of growing; which is not only a disappointment to the purchasers, but likewise an injury to the practical gardeners' – words that Fairchild would echo in *The City Gardener*.

In 1701 the company sought to expand its activities to cover the whole country, not just London, so that unskilled practitioners anywhere would be unable to escape its jurisdiction, and more young men could serve legitimate apprenticeships. A proposal presented to Parliament on behalf of the Master suggested that the company should be allowed to recruit into its ranks another 560 members – 10 garden designers, 150 gardeners to noblemen and 400 gardeners to gentlemen. Again, nothing came of the initiative; but in any case the livery companies were by the early eighteenth century beginning to lose their power to control London's trade and economy. Peter Earle writes in *The Making of the English Middle Class* that they were fast becoming 'the wealthy dining clubs with important charitable functions which most of them are today'.

Nursery work was specifically included in the Gardeners' Company's definition of the 'mystery of gardening', alongside potagery (kitchen gardening), florilege (flowers), orangery (greenhouse or conservatory work), sylvia (planting woods and lawns), botany and garden design. In his book *The London Tradesman*, published in 1747, R. Campbell described gardening as 'a healthful, laborious, ingenious and profitable trade'. Nurserymen and owners of seed shops furnished gentlemen with young trees and plants. 'It is a very profitable branch and in few hands,' he noted, adding that successful nurserymen must be 'compleat gardeners'.

For his first dozen years or so at Hoxton, Fairchild did not apply for membership of the Company of Gardeners – maybe because he did not regard himself as yet 'compleat' but more likely because he was not the principal owner of the nursery. He certainly felt strongly about the need for adequate training for young gardeners. Bradley, in his *General Treatise of Husbandry and Gardening*, quoted Fairchild as remarking that gardening apprentices had less chance than those in any other trade of learning their craft thoroughly, for a seven-year apprenticeship covered only seven growing seasons, while 'in other arts the students in them may daily observe what they discuss'.

Bradley agreed with him and took the opportunity of heaping yet more praise on 'the most rational gardener I have ever met'. He wrote:

> The more I reason upon, the more I am persuaded he is in the right; but his veracity is already well known, as his skill is unquestionable. And indeed 'tis by conversing with such men who have a true bent of genius to their business, and time and opportunity for practice, that a curious man may receive instruction; and not to have a too great opinion of one's self, which is always an impregnable bar against reason, upon which the arts of husbandry and gardening chiefly depend.

Fairchild had touched on the theme of gardeners' training in his own book, and Bradley returned to it repeatedly. In another part of the *General Treatise* he suggested a reason why there were so many bad gardeners about:

> The country people generally pick out such of their children to employ in husbandry as they judge are not worthy of good education; and whom they suppose have so little genius that they are only fit to drudge in hard labour . . . Husbandry and gardening ought rather to fall under the care of expert philosophers and reasonable men, who have judgment enough to remark the different effects of different seasons; the situation of the lands they are to cultivate; the depth and quality of their souls.

He propounded a method whereby garden owners may distinguish between genuine gardeners and mountebanks: 'The chief motive which has prevented the gentlemen from the hire of gardeners has been that they have employ'd persons that were inaccurate in their judgment and thereby have had their plants destroyed, and that such gardeners who really deserved by their ingenuity the favour and esteem of the gentlemen, had not an opportunity of showing themselves in their true lights.' He proposed that the Gardeners' Company should organise a practical test and grant diplomas to those it found competent, 'which will prevent entirely those that are pretenders to the art from being employ'd, who commonly do more damage to a garden in one year than can be retriev'd in three'.

By 1704, when he was already 37, Fairchild felt a need to regularise his position with the guild. This supports the evidence of the rate books that he had been in control of the business since 1703. (If that was indeed the year he took it over, he could hardly have chosen a worse one. On the night of 26 November one of London's severest gales ever was recorded, with hundreds of buildings damaged and destroyed and hundreds of ships lost in the Thames. Gardens and nurseries would not have escaped its ravages. But he would have had time to put matters to rights before the start of the summer of 1704, which John Evelyn described in his diary as 'very plentiful'.)

Fairchild's status as a potential member of the Gardeners' Company was complicated by his having served an apprenticeship in a different trade, although this was not a unique circumstance and there were provisions in the regulations for it. The rule was that anyone qualified by servitude to be a member of another guild, yet now working as a gardener, should become a freeman of both guilds simultaneously – of the original guild by servitude and the Gardeners' Company by redemption, or payment. That, according to the records, is precisely what Fairchild did on 13 June 1704.

First he would have gone to the Clothworkers' Hall in Mincing Lane, rebuilt a few years after the Tudor hall on the site had been destroyed in the Great Fire of 1666. It was next to the Church of All Hallows Staining, which survived the fire, only to collapse five years afterwards; but that, too, had been rebuilt by 1704. (The tower of the old church is there today, alongside the latest Clothworkers' Hall, built in the 1950s.) He cast his expert eye over two mulberry trees that stood outside the hall: 18 years later, in *The City Gardener*, he described them as having 'stood there many years, and bear plentifully, and ripen very well'. Having accepted his freedom from the Clothworkers' Company, Fairchild went to pay his redemption fee to the Gardeners' Company, at whatever nearby hall, tavern or coffee shop they were using as a meeting place; for they have never had a hall of their own. Thus, he became freeman of two city guilds on the same day.

All this would have cost money. In *The London Tradesman*, Campbell estimated that between £500 and £1,000 would have been required to set up as a master nurseryman. It is unlikely that he would have been paid enough for working there in earlier years to have amassed that sum. This could have been the point at which he sold the rights to the copyhold in Aldbourne that he had inherited from his father.

It was an auspicious year to be accepted formally into the City fraternity. On 13 August 1704, exactly two months after Fairchild became a freeman, the Duke of Marlborough's army beat the French at Blenheim, and on 7 September there were massive celebrations, highlighted by the presentation of a ceremonial sword to Queen Anne at Temple Bar, the City's western entrance. John Evelyn reported: 'Every company was ranged under its banners . . . Music and trumpets at every City company.' After a service at St Paul's Cathedral, 'the City companies feasted all the nobility and bishops, and illuminated at night'. Fairchild, newly admitted to two of the companies, would surely not have wanted to miss the party.

He does not appear to have kept up his links with the Clothworkers' Company after his induction, but he was an active, even a militant member of the Gardeners' Company, and as he learned more about it he began to share the indignation of many members both about the way it was run and the low status afforded it by the City fathers. In January 1721 several freemen petitioned the Lord Mayor and the City aldermen to look into the activities of the company's Master and wardens who, according to the petition, had neglected to enforce the rules protecting their members' monopoly on the trade. As a result, 'all the evils are crept in for the prevention of which the said grant [of the charter] was made, to the prejudice of His Majesty's subjects as well as the ruin of your petitioners and their families'.

The original document does not survive, so it cannot be said for certain whether Fairchild was one of the signatories. It is probable that he was, given an incident the following year in which he was certainly involved. He was one of seven members of the Gardeners' Company to try to exercise a vote in the Parliamentary election of April 1722, to elect four MPs to represent the City. The vote was restricted to the 8,500 liverymen (the number had doubled since the Restoration) in the 61 companies whose livery was recognised by the City authorities. As the gardeners' livery had never been recognised, the votes of the seven men were certain to be discounted and they would be listed on the returns as 'bad pollers'.

Voting irregularities were fairly common in the eighteenth century, and on 17 April the Lord Mayor, having seen the election records, summoned the clerks of the livery companies to submit their lists of liverymen to be checked against the ballot record. A week later, still dissatisfied, he called them all in again, and as a result of his objections just over a thousand votes were declared invalid, from people who were retired from their guilds or were not on approved lists for some other reason, or in some cases had voted twice. The twenty-one votes from the seven gardeners were entered in a separate category of unacceptability, and, since the ballot was not secret, the poll book records who they voted for.

Although the City commanded four seats in Parliament, most of the gardeners voted for only three of the six candidates who presented themselves. The 'ticket' that they supported, with only slight variations, was Humphrey Parsons, Francis Child and Richard Lockwood. The last two were elected but not Parsons, a brewer who never became an MP but later was twice Lord Mayor of London. Fairchild had in fact broken ranks with the others and did not vote for Parsons, instead plumping for John Barnard, another of the successful candidates, about to start a 39-year stint as an MP for the City. Barnard, born a Quaker, would become one of the most influential figures in eighteenth-century London, and a statue of him was put up in the Royal Exchange during his lifetime. In 1732 he was knighted by George II, on the same occasion as his fellow City MP Francis Child whose 1722 victory was also his first.

All seven gardeners voted for Child, head of the banking firm founded by his father. The winning candidates were all to some extent opponents of Sir Robert Walpole's Whig administration, which existed in a constant state of antagonism with the City, each side jealous of the other's powers and influence and anxious to protect its own. Although Barnard technically ran as a Whig in this election, he made himself a nuisance to Walpole and ran as a Tory populist next time.

The seven gardeners must have known that their votes would be discarded when the authorities came to check the names of voters against the membership rolls of the recognised companies. In making the gesture – and in casting their invalid votes for popular City champions – they were expressing their solidarity with freemen of other guilds and at the same time highlighting the discrimination against their company. It was not until 1891 that the livery of the Gardeners' Company was finally recognised by the City aldermen.

By 1722 Fairchild was London's leading specialist nurseryman. His book *The City Gardener*, published that year, was a unique and valuable handbook of plants that could be grown in the rapidly expanding and increasingly polluted metropolis, a guide to urban planting both in public squares and private gardens, formidably detailed but at the same time still highly readable today.

Considering that gardening had not long been established as a popular pastime, there had already been a fair number of books devoted to it, and they were appearing at an increasing rate. Blanche Henrey, in her formidable work *British Botanical and Horticultural Literature before 1800*, notes that, of the 100 gardening books published in the seventeenth century, 80 appeared after 1650. The total rose to 600 in the eighteenth century, when publishers began to recognise that keen gardeners were at least as fond of reading about their passion as they were of actually getting their hands dirty – a preference that persists to this day, if we are to judge by the generous amount of space that bookshops allot to manuals and glossy illustrated works on the subject.

Most early gardening books were herbals, essentially lists of herbs and flowers with notes of the medicinal qualities attributed to them – some actual, some supposed. The first was *The Great Herbal*, a translation from a French work that appeared originally in 1516. Sometimes, as in Nicholas Culpeper's *Complete Herbal* of 1652 and John Archer's *Compendius Herbal* of 1673, astrological instructions were added, advising on the most favourable phases of the moon for sowing and harvesting to allow the juices to work most effectively. They included basic descriptions of plants, chiefly for the purpose of identification, for before Linnaeus there was great confusion about their naming. Culpeper uses words like madwort, sourweed, ladies' smock, hare's ear and hound's tongue, which sometimes referred to different plants in different regions of the country.

There were some more practical manuals. John Worlidge's *Systema Horticulturae*, also published in 1673, is sometimes cited as the first book of general instruction for gardeners, although a treatise on the subject by Thomas Hill survives from as early as 1560, and Richard Gardiner's *Profitable Instructions for the Manuring, Sowing and Planting of Kitchen Gardens* appeared in 1603. Gervase Markham wrote a number of books in the early years of the seventeenth century, including *The English Arcadia*, *The English Husbandman*, *Country Contentments* and *The Expert Gardener* – published as one volume in 1640, three years after his death.

The cultivation of trees, both for fruit and timber, became fashionable during Cromwell's Commonwealth: Samuel Hartlib wrote *A Design for Plentie by an Universal Planting of Fruit Trees* in 1652, and a year later came Ralph Austen's *Treatise of Fruit Trees*. John Evelyn's celebrated *Sylva*, a paean to trees, was published in 1664, and so was his *Kalendarium Hortense*. The same year saw the appearance of Stephen Blake's *The Compleat Gardener's Practice*. John Ray produced the first systematic English flora in 1670 and the first volume of his ground-breaking *Historia Plantarum* in 1686. Leonard Meager's *The English Gardener* was published in 1678 and his *New Art of Gardening* in 1697.

Nehemiah Grew's scholarly *Anatomy of Plants*, in which he first hinted at the possibility of sexual reproduction of flowers, appeared in 1682, and six years later came the first recorded nurseryman's plant catalogue, issued by George Ricketts. The nurserymen George London and Henry Wise published *The Retir'd Gardener*, a translation from the French, in 1706. Stephen Switzer, a celebrated gardener and garden designer, wrote *The Nobleman, Gentleman and Gardener's Recreation* in 1715, and the following year came the first of Richard Bradley's many and various books on the subject.

It was Bradley, according to his own evidence, who persuaded Fairchild to burst into print and presumably helped arrange matters with the joint publishers, Thomas Woodward of St Dunstan's in Fleet Street and John Peele of Paternoster Row. In his monthly *General Treatise of Husbandry and Gardening* for August 1621 Bradley published a letter he had written to a friend saying: 'I have at length persuaded Mr Fairchild to publish his remarks on the London gardens.'

In contrast to Bradley and other gardening authors, Fairchild was targeting a new market. *The City Gardener* was the first gardening book not directed primarily at owners of country estates. Responding to the growing fashion for creating rustic oases in heavily populated areas, the book's principal aim was to help this new band of enthusiasts develop horticultural skills and in particular to make the best of the prevailing conditions in the City – not always conducive to successful husbandry.

In the late seventeenth and early eighteenth centuries the most insidious blight of London's day-to-day life was the increasingly foul atmosphere. The prolific John Evelyn wrote a pamphlet called *Fumifugium, or the Inconvenience of the Air and Smoke of London Dissipated*. Coal was being brought in by the shipload and used in ever greater volume in hearths for domestic heating and in furnaces for industries, such as the flourishing potteries by the Thames. It was known as sea-coal to distinguish it from charcoal – either because it arrived in London by sea from Newcastle upon Tyne, or because in early times it was picked up from the Northumberland beaches. (*The Oxford English Dictionary* prefers the latter explanation, pointing out that early uses of the term are not confined to London or other ports.)

People had long recognised that coal smoke and dust must be bad for them, as well as unpleasant to the eyes and nose. As far back as 1257 Eleanor, Henry III's queen, was forced to leave smoky Nottingham, where she was staying in her husband's absence, and move to a more rural castle. From the thirteenth century onwards, several attempts were made to prohibit the burning of sea-coal in London, and one man is said to have been executed for continuing the practice after being warned. If that is true, he was unlucky to be singled out, for coal-burning seems to have continued with little interference. In 1648 a group of citizens, we would now call them environmentalists – urged Parliament to ban the importation of the foul black stuff into the capital – but without success. Perhaps their failure was a long-term blessing, for if they had prevailed we might have been deprived of Christopher Wren's masterpiece, St Paul's Cathedral, whose construction after the 1666 fire was funded by a tax on sea-coal.

Between 1700 and 1722, when *The City Gardener* was published, the amount of coal imported into London annually rose by a third, from 335,000 chaldrons to 460,000. (A 'chaldron' was a measure used only for coal, equivalent to 32 bushels.) Today the trade is recalled in the name of Old Seacoal Lane, north of Ludgate Circus, where the vessels from Newcastle would unload their cargoes after turning north from the Thames into the Fleet River.

Smoke affects plants in two main ways. The emissions of sulphur dioxide can cause organic damage, and the thick dark clouds prevent them from absorbing the amount of light they need to prosper, especially in winter. Today we know that smoke can also have its beneficial effects, mainly to hinder the growth of black spot, the fungal disease that afflicts roses: ever since coal-burning was ended in London in the second half of the twentieth century, black spot has been on the increase.

Fairchild may not have understood the science, but he observed that some plants did not thrive in the smoky conditions, performing far worse than in the clear air of his native Wiltshire. 'In places in London where every part is encompassed with smoke, and the air is suffocated or wants its true freedom,' he wrote in *The City Gardener*, 'plants, which generally are used to the open air, will not be always healthful: and therefore I have made it my business to consult what plants will live even in the worst air of chimneys.'

A passage on pruning evergreens is a good example of how he thought through his techniques of managing plants in this polluted environment, based on acute observation and by making analogies – perhaps inspired by his medical friends – between the treatment of plants and the treatment of people:

> The evergreens which I have mentioned do not thrive so well with much cutting as they will do otherwise; for the smoky air of the town seems to have a very considerable effect on them when they are pruned, though it is still convenient to prune off the dead wood when we find it. We must consider that in nature there is no such thing as pruning; and when a tree is under the power of the London smoke, which is not so free and open nor so healthful to it as the country air, it has enough to do to support life; and it would therefore do it a double injury to wound it with the knife, when it wanted convenient help to heal its wounds and was but low in health . . .
>
> Some learned men say that whatever can be made agreeable to a sick man will help his cure or contribute to his health; but whatever is the contrary increases his distemper and might even cause his death: and sure nothing could be more tending to his detriment than wounding him when his body was already weak and low . . . In a word, 'tis not every tree that will grow in London that will bear pruning.

On the very first page of the book he declares that the acrid smoke from the chimneys is among the principal scourges of the metropolitan gardener, along with the universal human factor:

THE

City *GARDENER*.

INTRODUCTION.

HAVE upwards of thirty Years been placed near *London*, on a Spot of Ground, where I have raiſed ſeveral thouſand Plants, both from foreign Countries, and of the *Engliſh* Growth; and in that Time, and from the Obſervations I have made in the *London* Practice of Gardening, I find that every thing will not proſper in *London*; either becauſe the Smoke of the Sea-Coal does hurt to ſome Plants,

or

The first page of Fairchild's *The City Gardener* (*Royal Horticultural Society, Lindley Library*)

> I find that everything will not prosper in London, either because the smoke of the sea-coal does hurt to some plants, or because those people who have little gardens in London do not know how to manage their plants when they have got them . . . I have therefore been advised to give my thoughts in this manner, that everyone in London, or other cities where much sea-coal in burnt, may delight themselves in gardening, though they have never so little room, and prepare their understanding to enjoy the country, when their trade and industry has given them riches enough to retire from business.

That, then, was his straightforward reason for writing the book; but the circumstances of its publication, like much else about him, contain an element of mystery. Most books at the time were dedicated in florid terms to wealthy patrons who had either subscribed to them or otherwise helped fund their publication, or from whom favours might be expected in the future. Fairchild's dedication reads: 'To the Governors of the Hospitals of Bethlem and Bridewell, these papers, for the improvement of London gardens, are most humbly presented and dedicated by their most obedient humble servant, Thomas Fairchild.'

Bethlem was London's asylum for the mentally disturbed, then in Moorfields, between Hoxton and the City, housed in a splendid building designed in the 1670s by Robert Hooke and overlooking a magnificent public garden. Bridewell was a Tudor royal palace on the Fleet River, near today's Blackfriars Bridge. Since the mid-sixteenth century it had been used both as a prison and an orphanage, supported by donations from the guilds and other sources, where poor boys were given a smart blue uniform and

To the Honourable and Worſhipful

The GOVERNORS

Of the HOSPITALS of

BETHLEM

AND

BRIDEWELL,

THESE

PAPERS,

For the Improvement of the

London GARDENS,

Are moſt humbly Preſented

and Dedicated by

Their moſt Obedient

Humble Servant,

Tho. Fairchild.

Dedication page of *The City Gardener* (*Royal Horticultural Society, Lindley Library*) taught skills that could allow them to become apprentices. It was badly damaged in the 1666 Great Fire and rebuilt. Since the sixteenth century the word Bridewell had also come to be used as a generic name for workhouses, and there was one in Clerkenwell.

In the absence of many solid facts about Fairchild's childhood and domestic life, it is tempting to read into these dedications more than is probably there. It has never been established for certain that Fairchild married, but there is a teasing hint in his will, where he makes a legacy to Mary Price 'my daughter-in-law'. Logic suggests that a man with a daughter-in-law must have been married, even if the word is interpreted as meaning a step-daughter, an alternative usage at the time. However, no mention of a wife is made in any contemporary references to Fairchild, or on his tombstone. He certainly had no direct heir, because he left his nursery to his young nephew Stephen Bacon, his mother's grandson from her marriage to John Bacon following the death of Fairchild's father.

A daughter-in-law is hard to explain without a wife. Had there, then, perhaps been some mental illness that led his wife to be confined to Bethlem, through which Fairchild became acquainted with the institution, and grateful to it? As for Bridewell, since so little is known of when he left Aldbourne to come to London, it is conceivable that his family exploited a link with the Clothworkers' Guild to have him taken in there, even though he was not an orphan. That his cousin John was apprenticed to the Clothworkers' Guild a few years later could suggest such a connection, and some of the Bridewell boys are known to have been apprenticed to outside masters.

These are rich fields for speculation, but a more likely

reason for the dedication lies in the text of the book itself. Fairchild clearly identified himself with the common herd and was keen to promote public gardens that could be enjoyed by all. There are references in the book to the gardens at Bridewell and Bethlem that show he was familiar with them and may have supplied some plants for them.

Everything suggests that he had a social conscience. In his will he left a little money to the newly created charity schools: the one at Shoreditch was among the first in London. In the fashion of the time, he was also conventionally pious and a practical moralist. In the book's introduction he offers words of comfort to those who can run to gardens of only limited size: 'True care and industry will make their gardens larger, as the same care will increase their fortunes.' It is clear from his other opening remarks that a direct relationship between gardens and fortunes had been forged in recent years, and gardening was fast catching on among the newly rich merchants of the City, as well as the landed gentry:

> One may guess at the general love my fellow citizens have for gardening, in the midst of their toil and labour, by observing how much use they make of every favourable glance of the sun to come abroad, and of their furnishing their rooms or chambers with basins [pots] of flowers and bough-pots [vases], rather than not have something of a garden before them. Nor is this pleasure less cultivated among persons of quality, while public affairs oblige them to the town during the busy days of the week; I have heard some say that the sight of good flowers, and their grateful smell, has made them often wish to be enjoying the pleasures of their country gardens. And so I find that

> men of business are all upon the same foot in seeking country pleasures.

And he goes on to underline his moral purpose:

> Now, when gardening goes so far among men in general as to engage the minds of the most worthy part of mankind, or I might say of all men who have the least time for diversion; I see no reason why I should not cultivate this innocent pleasure among my fellow citizens; that from the highest to the lowest, every one may be improving their talent, or even their mite, in the best way they can, in order to increase their quiet of mind, to be fixed in a right notion of country happiness, when their affairs will permit them to reach such pleasures.

His confident style, opinionated but not brash, is of a piece with the practical, down-to-earth qualities that attracted his scientific friends and led them to sing his praises in their accounts of his experiments. Were he living today, he would have been a perfect television gardener, swarthy and weatherblown, dispensing sound and authoritative advice to experts and novices alike, without talking down to any of them: the Geoff Hamilton of Georgian London.

In the absence of any personal diaries or documents, *The City Gardener* provides our best first-hand clues to the circles Fairchild moved in and the range of his acquaintances. In the book he mentions the gardens of most of the grand London houses of the aristocracy, which he probably visited for professional rather than social reasons. His many references to plants growing in specific parts of London show that he walked all over the City to see how its gardeners were creating their little patches of urban rusticity. A vivid picture of these green oases can be conjured by devising a route that links these references, using the book as a guide to the flora that he and his contemporaries observed there. The exercise involves some cheating with the seasons, for not everything he describes would be in flower at the same time, but it is instructive none the less.

We start at Crutched Friars, the oddly named street that today runs beneath Fenchurch Street Station. In a courtyard here Fairchild noticed a thriving bladder senna (*Colutea arborescens*), a tall, yellow-flowered member of the pea family, seldom grown nowadays. From here we descend Tower Hill, passing the tower of All Hallows, Barking, from where Samuel Pepys viewed the destroyed City after the 1666 fire. Beyond it, the Tower of London then marked the south-eastern edge of the City. There, by the Record Office, were apple trees in pots, grafted on to paradise stock – the name given to the dwarfing stock then in most common use.

Fairchild recommended these apples mainly for the beauty of their spring blossoms and noted that at the Tower, ventilated by river breezes, they had blossomed for five successive years without being moved from their position. In more built-up sections of the metropolis he suggested letting them overwinter in the clean air of the country and bringing them to London when the blossoms broke.

From the Tower we walk west along Thames Street. On our left the narrow passages and alleys run down to the busy quayside, where scores of sailing ships are tied up loading and unloading goods, with many others anchored

This map shows the City as Fairchild would have known it in the 1720s. The Stocks Market is at lower right, with the Guildhall to its north-west and Moorfields at top right, outside the wall. *(Guildhall Library, Corporation of London)*

in the river awaiting a berth. The centre of activity is the Custom House, a fine Wren building damaged in a gunpowder explosion in 1714 and rebuilt over the next ten years to the design of Thomas Ripley. (Today's building dates from 1825.) Fairchild would have been known here: it is where he had to go to collect the plants that Bradley sent him from the Netherlands and Catesby from America – probably a lengthy and quite expensive procedure, involving the greasing of various eager palms.

On the other side of Thames Street, up Idol Lane, is the church of St Dunstan-in-the-East, its tower and spire rebuilt by Wren after the Great Fire. Nearly 300 years later, after bombing in the Second World War, the tower is all that remains of the church, standing in a small garden of which Fairchild would certainly have approved. In his book he mentions the plane trees that prospered in this churchyard and had grown to 40 feet high. This is impressive, because the London plane had been introduced to Britain from southern Europe only some 40 years earlier. Today, no planes remain in the church garden, but there are oaks, beeches and magnolias, along with shrubs that include camellias, potentilla, lavatera, laurels, hypericum and fuchsia.

Rejoining Fairchild on Thames Street, with the bustle and smell of Billingsgate fish market on our left, we cross the approach to the medieval London Bridge, still lined with the old houses that in 40 years would be demolished to make room for increasing traffic. Passing the church of St Magnus the Martyr, we look right to catch a glimpse of Wren's Monument, commemorating the Great Fire. Just beyond it a few grand houses of City merchants still stood in Fairchild's time, although they would soon be subsumed by the area's increasingly commercial character.

Now we turn right up Queen Street, built after the fire as an approach to the Guildhall, following the line of the narrower Soper Lane, where soapmakers used to trade. At the Stocks Market (a meat and fish market named after the pair of stocks that stood in the thirteenth century on the site of today's Mansion House) we turn left, passing the south face of St Paul's and then the old city gate in Ludgate Street, the present-day Ludgate Hill. Fairchild tells us that here, outside Sam's Coffee House, two large mulberry trees grew in a yard about 16 feet square.

We cross the filthy Fleet Ditch, a notoriously unhygienic canal leading from the Fleet River to the Thames, broadened by Wren after the Great Fire but in Fairchild's time blocked again with foul garbage, until it was finally covered over in 1733. The Fleet Bridge that crossed it was, by contrast, quite pleasant, decorated with pineapples and the City of London's coat of arms. The area around here was mostly built up, and Fairchild marvelled that fig trees prospered 'although encompassed with houses on every side which are so high that the sun never reaches them in winter'. The trees were some 15 feet tall and carried leaves from near the ground to the top, 'for which reasons I wonder it has not been more generally propagated in the city gardens, especially since they will not only thrive well in London but bear good fruit too'.

We are making for Bridewell, which despite its situation boasts its own garden. Fairchild observes here a mezereon (*Daphne mezereum*) with striking purple flowers, as well as a thriving privet, covering walls six to eight feet high, 'which is best pruned in winter', he advises in passing. From there we walk along the Thames, skirting St Bride's church, until we reach the gardens of the Temple, which Fairchild refers to often. This was where, at the start of the Wars of the Roses, supporters of the Yorkists and Lancastrians are said to have plucked their respective white and red roses, which became their county emblems.

Nearly 300 years later, among trees that Fairchild notes here are elms, horse chestnuts (in the Master of the Temple's garden, where there is little or no sun) and a vine in the courtyard of the Rose Tavern, also starved of sunlight. The garden is well provided with flowers – wallflowers, stocks, carnations and 'a good number of exotic plants'. The Master from 1704 to 1753 was Dr Thomas Sherlock, later Bishop of London.

Continuing along the Strand past Twining's tea shop (its descendent is still there) and James Gibbs' newly built church of St Mary, we turn towards the river into Norfolk Street (between the present Arundel and Surrey Streets), where the town house of the Dukes of Norfolk stood until 1678. We pay a call on Mr Jobber, 'a very curious gentleman' and obviously a keen gardener, with a special interest in cacti and succulents. Fairchild admired his aloes and 'those strange plants called torch thistles' (a kind of cactus) and marvels at the excellent Jobber's ability to keep them going in winter, advising that they should not be watered between Michaelmas and May. More conventionally, Jobber also produced a good summer display of fig marigolds, which we know today as mesembryanthemums.

We reach Somerset House, the Renaissance palace built in the mid-sixteenth century, which in 1775 would be pulled down and replaced by the imposing office building that stands there today. Fairchild tells us that this garden has 'been observed to produce several varieties of things, which the more inland parts of town have not generally been garnished with', convincing him that the air by the river is less polluted, thus more flower-friendly, than that further inland. Another contemporary described three stately avenues of trees in the Somerset House garden.

Now we pass the Savoy, the medieval palace used in the eighteenth century as lodgings and workshops before being pulled down in 1816 to make way for the approach to Waterloo Bridge. Beyond where the Savoy Hotel stands today was a group of late seventeenth-century town houses built by the speculator Nicholas Barbon on the grounds of the former 'Strand palaces' – homes of the aristocracy that were mostly abandoned soon after the Restoration. A few lasted longer, including our next landmark, Northumberland House, the imposing Jacobean residence of the Percy family that would survive until 1874.

In Whitehall we peer at the late Mr Heymen's garden, which boasts 'several pots of flowers, both auriculas and carnations, which blossomed very well'. The gardens of the old Whitehall Palace, neglected since the palace was burned down in 1691, are said to have been at their best in the reign of Queen Elizabeth. In a narrow passage between New and Old Palace Yards at Westminster, Fairchild admired a Virginian acacia (*Robinia pseudoacacia*, the locust tree). And in the Earl of Halifax's garden near Parliament House, he noted more plants that survived in town only because they benefited from the river's breezes.

We head away from the Thames to cross St James's Park, one of the most pleasing green spaces in London:

> Some gentlemen, who have been abroad, have told me that there is no public place for walking in any city this side of Italy that is so pleasant as St James's Park. The gardens belonging to the French king at Paris are not near to it in beauty, as I am informed.

From his description it is apparent that the park, though remodelled in the intervening 280 years, has retained its essential character:

> The Park at St James's is of a large extent and disposed in handsome walks of lime trees and elms, a large regular canal, a decoy for ducks. And although it is as

> much oppressed with the London smoke as almost any of our great squares, yet the wild fowl, such as ducks and geese, are comfortable to it and breed there; and there is an agreeable beauty in the whole that is wanting in many country places.

His egalitarian instincts lead him to suggest, perhaps a bit mischievously, that such delights might be replicated in the less glamorous parts of London close to his nursery: 'The quantity of ground which now lies in a manner waste in Moorfields might undoubtedly be rendered very agreeable were it to be adorned after the same manner, and be as delightful to the citizens as St James's Park is to the courtiers.'

Adjoining St James's Park were the gardens of some of the most prominent of those courtiers: the recently built Buckingham House (later Buckingham Palace), Godolphin (now Lancaster) House and Marlborough House, as well as the private garden of St James's Palace. Fairchild notes that they all produce good fruit and flowers, and that 'the common birds of the woods are familiar in these gardens, as well as the park'. His expertise may not have extended to ornithology, for he makes no attempt to name the birds he was referring to.

Buckingham House was built in 1705 for John Sheffield, the first Duke of Buckingham, who described the garden in a letter in 1709. A gravel walk had a covered arbour at each end, two groves of limes were bordered with bay and orange trees, and a further double row of limes lined a canal 600 yards long. There was a terrace 'from where are beheld the Queen's two parks and a great part of Surrey', and on one side of it 'a wall covered with roses and jasmine is made low to admit the view of a meadow full of cattle just beneath (no disagreeable object in the midst of a great city)'.

At each end of the terrace were parterres with fountains, one of which led on to a square garden with another fountain in the middle and greenhouses at the side, and a kitchen garden below it. Fairchild's observations about the birds were confirmed by the duke, more specifically: 'Under the windows of this greenhouse is a little wilderness full of blackbirds and nightingales' – foreshadowing the modern taste for wildlife gardens alongside more tightly designed spaces.

On, then, up Pall Mall, a street of more grand houses and fashionable shops. As John Gay wrote in 1716:

> Oh bear me to the paths of fair Pell-Mell!
> Safe are thy pavements, grateful is thy smell . . .
> Shops breathe perfumes, through sashes ribbons glow
> The mutual arms of ladies and the beaux.

Dodging these mutual arms we turn into St James's Square, of which Fairchild was not overly fond. It was mainly cobbled, with a post-and-rail fence around it. He did not like the formal layout of this and similar London squares created in the second half of the seventeenth century:

> The plain way of laying out squares and grass platts [small flat lawns] and gravel walks does not sufficiently give our thoughts an opportunity of country amusements. I think some form of wilderness-work will do much better, and divert the gentry better than looking out of their windows upon an open figure.

Members of Parliament may well have been influenced by these strictures, because in 1725 they appointed trustees to improve the square, although shrubs and trees were not planted until the following century.

The book contains several sketches of possible layouts for city parks – none of them, it has to be said, especially inspired. Given that he was not averse to self-advertisement, he may have hoped to embark on a second career as a garden designer, creating the beds for which his nursery would provide the flowers. There is no evidence, though, that any potential patron rose to the bait.

In expressing his preference for natural over formal garden design, Fairchild was slightly ahead of his time. It would be another 30 years before Capability Brown began ripping up the elaborate parterres from scores of country houses and replacing them with his skilfully contrived

A suggested plan for a formal public garden from *The City Gardener* (*Royal Horticultural Society, Lindley Library*)

rustic landscapes. On a smaller scale, that is what Fairchild advocated for London squares. He might well approve of St James's Square as it enters the twenty-first century, for although the garden still has its symmetrical paths and beds there is also some quite thick planting of trees and shrubs near the edges: not exactly a wilderness, but it avoids the bare, cold formality that Fairchild found offensive.

Into Haymarket (hay and straw continued to be sold there until the 1830s), where he certainly approves of a fountain outside a plumber's shop at the Piccadilly end. He is keen on water features and uses this one as an example of how they can successfully be incorporated into urban landscapes. In nearby Leicester Fields (Leicester Square) he notes a vine that bears good grapes every year and remarks that vines also do well in tavern yards and other enclosed places. The district of Finsbury, just north of the City, is said originally to have been called Vinesbury.

Eighteenth-century Soho was already a cosmopolitan area, crammed with the apartments and workshops of hundreds of French Huguenots who came to London after the revocation of the Edict of Nantes in 1685. At its centre, though, Soho Square, laid out at around the same time, was an oasis of splendid town houses. Contemporary prints show the centre of the square as an area of symmetrical lawns of the kind Fairchild disliked, but there was a smattering of shrubs in it. One he notes with satisfaction is the syringa, now a generic name for lilac but then used for the philadelphus, or mock orange – also called the pipe tree because its stems were used to make pipes. The statue of Charles II that dominated the square is still there today, but without its tall plinth.

North-east to Bloomsbury Square, also filled with fashionable housing, where Fairchild notes another *Robinia pseudoacacia*: there are two of them in the square today, but certainly not the same ones. Close by in Lincoln's Inn Fields – then none too savoury a place – he spots a little enclave of yellow flowers: another bladder senna, the somewhat similar Scorpion senna (*Coronilla emerus*) and the familiar *Cytisus* (broom). Continuing east we reach Chancery Lane and the Rolls Chapel – part chapel and part legal record office, rebuilt by Inigo Jones in 1617 – in whose garden 'figs have ripened very well'.

We are now walking through legal London. Our next horticultural landmark is a coffee house in High Holborn by the gate to Gray's Inn, one of the Inns of Court where barristers are trained. Here Fairchild, as a skilled grape cultivator, is intrigued to see 'a vine which grows very well in a small pot, though it is constantly kept in a close room; this year it was full of leaves before Christmas'.

Crossing Gray's Inn gardens to leave on the north side, we have quite a long walk to our next port of call. Clerkenwell Road will not be constructed until the 1870s, so we have to find our way past Hatton Garden, back across the Fleet River to Clerkenwell Green, passing the remains of St John's Priory to enter the streets of Clerkenwell, a mixture of solid merchants' houses and workshops for Huguenot clockmakers and other craftsmen. We are making for Charterhouse, a former monastery that became a school and almshouses in 1614. The three-acre garden here was more to Fairchild's taste, for it had a 'wilderness' element, and twice in his book he mentions the splendour of its white thorns (hawthorn).

In 1724 he was to become quite closely involved with Charterhouse, as one of five prominent experts invited by the head gardener, George Field, to make a report on the garden to the governors of the institution. Field's main purpose in commissioning the report seems to have been to convince his employers that he was overworked and underpaid. Fairchild and the other four clearly agreed. Their report suggested that at least two more gardeners were needed to help Field, especially since the kitchen garden was not nearly big enough to supply all the vegetables needed by the Master of Charterhouse and the 40 pensioners and would have to be expanded. There followed a detailed list of the cost of the recommended changes, amounting to £31.6d.

No action appears to have been taken, for in November followed a petition to the governors from Field asking for a pay increase to £30 a year. This was accompanied by a further document, signed by Fairchild and three others (including two who had not been involved in the report), affirming that the garden could not be maintained properly for less than £71.4s. a year – whereas the present budget, including Field's salary, was only £21.6s.8d. It is not recorded whether the governors accepted the findings of Fairchild and his brother gardeners: management consultants stand a better chance of having their recommendations implemented if they are appointed by the management, not the workers.

Leaving Charterhouse, we reach Cripplegate, one of the northern gates of the City (demolished in 1760), where Fairchild had given some advice on pruning figs to the Revd Dr Thomas Bennet, the parish priest at St Giles, described as 'a tall, strong and haughty man' and noted for his fervent denunciation of Quakers and other dissenters. Fairchild was confident of the beneficial effect of 'the new way of pruning' and mentions it twice in his book, without saying exactly what it involved. Having examined the figs after the pruning, he appeared to have been infected with

Dr Bennet's dogmatism, as he predicted: 'I question not that they will ripen well'.

There are more fruitful delights in alleys off the Barbican and Aldersgate Street, and further east in Bishopsgate Street, where Fairchild reports that pears grow well: 'Besides the fine show they make when they are in flower, they will bear very good.' This is surprising, as pears generally demand more sunlight than apples and are less tolerant of frost. They also have to be established for several years before they bear fruit ('pears for your heirs' is the orchardist's motto), although after that initial period they can remain productive for several generations.

Bishopsgate in the early eighteenth century was still the site of the homes of wealthy merchants, who would begin to migrate towards the West End only in the middle of the century. Some of the great Elizabethan houses survived there in Fairchild's time, their gardens surrounded by walls too high for pedestrians to peer over, and providing good protection for tender fruits and flowers.

We walk south, passing through London Wall at the south-east corner of Moorfields – the open space that Fairchild would like to see laid out like St James's Park. Here is Robert Hooke's splendid Bethlem Hospital – to whose governors Fairchild dedicated *The City Gardener* – and just inside the city wall the extensive garden of the Drapers' Company. This was one of the finest guild gardens in London, and the company's records show that large sums of money were spent on it in the first half of the eighteenth century, principally on walls, statuary and paving rather than on plants. It was open to the public, but from time to time restrictions were imposed to keep out such perceived enemies of horticulture as children, nursemaids, Jews, servants and gamblers. In 1715 the company court decided to close it on Saturdays, the Jewish sabbath.

Now we are in guild territory and pass many impressive halls – mostly rebuilt after the 1666 fire – until we reach the Guildhall itself, which had two gardens. There, or more precisely in 'a close place at the back', Fairchild wants us to notice a lily of the valley, flowering well every year. As at Bridewell, privet did well here and in the adjoining street, Aldermanbury, where he also spotted a fraxinella (*Dictamnus albus*, or burning bush), with its spikes of white flowers and scented pale green leaves. An apothecary in Aldermanbury, Mr Smith, boasted a fine collection of 'succulent and juicy' aloes.

This has been a long walk, and Fairchild would surely be keen to get back to his tender charges at Hoxton, so we turn north from the Guildhall, back through the city wall and up Grub Street, which the seventeenth-century poet Andrew Marvell pinpointed as a notorious haunt of jobbing writers. (Unhappily, its name was changed to the classier but less appealing Milton Street in 1830.) Just to its east, at Whitecross Street, Fairchild was gratified to find hops 'growing very vigorously in a closed alley', although probably not in sufficient quantity to supply the brewers in the City and East End. In another alley, between Whitecross Street and Bunhill Fields – the burial ground for dissenters – he spotted a further white thorn.

After crossing Old Street, it is about half a mile's walk to Hoxton, where perhaps he would have dropped in to exchange gardening chat with his friend and neighbour Benjamin Whitmill, whose winter-sown French marigolds he mentioned favourably in the book. Back at his own nursery, he will have discovered how many readers of *The City Gardener* had responded to the blatant commercial message in the last chapter:

> Perhaps many that have gardens in London are acquainted but with few sorts of the plants or flowers that I have set down for beautifying the city gardens. Their best way therefore to be informed will be to view the gardens at Hoxton and other places near the town, where they may see all the variety of flowers that blow [bloom] in the spring, summer and autumn season; and then consult with the gardener about those they like best, especially about which should be planted in autumn and which in spring.

If they had a garden of any size they would also need a jobbing gardener, and on this subject, too, Fairchild offered trenchant advice:

> The next thing to be considered is to have a gardener of judgment to manage a city garden; for a gardener that has been bred in the country, and has not had practice about the town, knows little more of managing a garden than one that is bred to plough and cart. There are many ignorant pretenders, who call at houses where they know there is any ground, let it be in season or out of season, and tell the owners it is a good time to dress and make up their gardens; and often impose on them that employ them by telling them everything will do, when perhaps it is a wrong season . . . This is a great discouragement, which makes those persons who delight a little in a garden neglect doing anything at all, thinking all their labour and cost thrown away.

Warming to the theme, Fairchild warns his readers of other charlatans bent on making money from the ignorant and unsuspecting. As we have seen, this was something that the Gardeners' Company had been exercised about for a long time. A century and a half earlier, in the 1570s, one of the earliest garden writers, Thomas Hill, had warned his readers against buying 'counterfeited' seeds that would not grow, or at least would produce different flowers and vegetables from those claimed by the vendor. Fairchild's target is traders who sell trees and plants in markets such as the Stocks. They are not nurserymen but fruiterers 'who understand no more of gardening than a gardener does the making up the compound medicines of an apothecary'. (Was this a sly dig at his friend Bradley and his Amsterdam adventure?)

> They often tell us the plants will prosper when there is no reason or hopes of their growing at all; for I and others have seen plants that were to be sold in the markets that were as uncertain of growth as a piece of Noah's ark would be, had we it here to plant. But when such plants are bought at the gardens where they were raised, there can be no deceit, without the gardener who sold them loses his character.

Interspersed with these advertisements for himself, Fairchild's book provides a revealing portrait of what Londoners grew in their gardens at the start of the age of horticultural enlightenment. It was clearly a limited range, and some of the plants he records with such enthusiasm are today regarded as little better than weeds. That is why he was captivated by collectors such as Mr Jobber of Norfolk Street: they shared his passion for the unusual, for the challenge of making plants prosper in the most unpromising conditions.

Curiosity and experimentation are the defining qualities of dedicated gardeners down the ages. Even today, when the range of available plants is many times greater than 300 years ago, the public thirsts for new varieties every season. Fairchild sought to slake that thirst, as much in the interest of business as of science, both by securing the services of plant-hunters and experimenting to produce new, pleasing flowers by crossbreeding. That was how he made his small piece of history.

Thomas Fairchild, painted by Richard van Bleeck in about 1723
(The Plant Sciences Library, University of Oxford)

Aldbourne church today. Fairchild may have been christened here and probably went to school in a room over the south porch *(Michael Leapman)*

Crane Court, eighteenth-century headquarters of the Royal Society *(By permission of the President and Council of the Royal Society)*

Fairchild's Mule, preserved in the Natural History Museum, London *(Natural History Museum, London)*

Carnations illustrated in *Paradisus Terrestris,* 1629 *(Royal Horticultural Society, Lindley Library)*

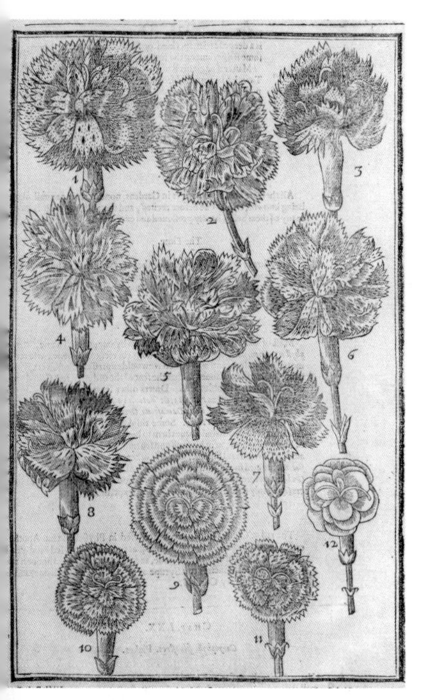

A modern variegated hybrid carnation *(Royal Horticultural Society, Lindley Library)*

Fairchild's white lily, from *The Compleat Florist,* 1740
(Royal Horticultural Society, Lindley Library)

Sir Hans Sloane, by Stephen Slaughter *(By courtesy of the National Portrait Gallery, London)*

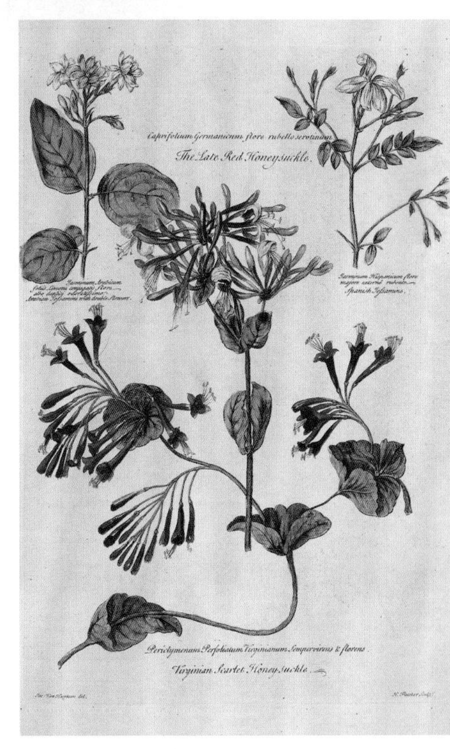

Honeysuckle, from *Catalogus Plantarum,* 1730 *(Royal Horticultural Society, Lindley Library)*

A tulip tree, sheltering a Baltimore bird, painted by Mark Catesby during his travels in Virginia *(Royal Horticultural Society, Lindley Library)*

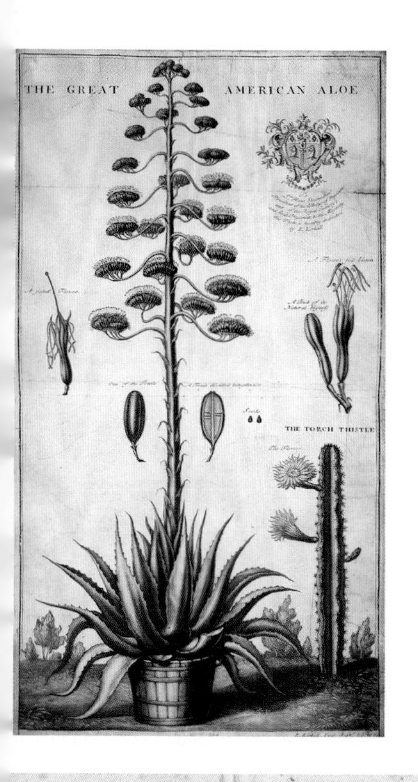

The great American aloe that bloomed in John Cowell's garden as Fairchild lay dying *(Natural History Museum, London)*

A thinly veiled advertisement for Fairchild's succulents from Patrick Blair's *Botanick Essays (Royal Horticultural Society, Lindley Library)*

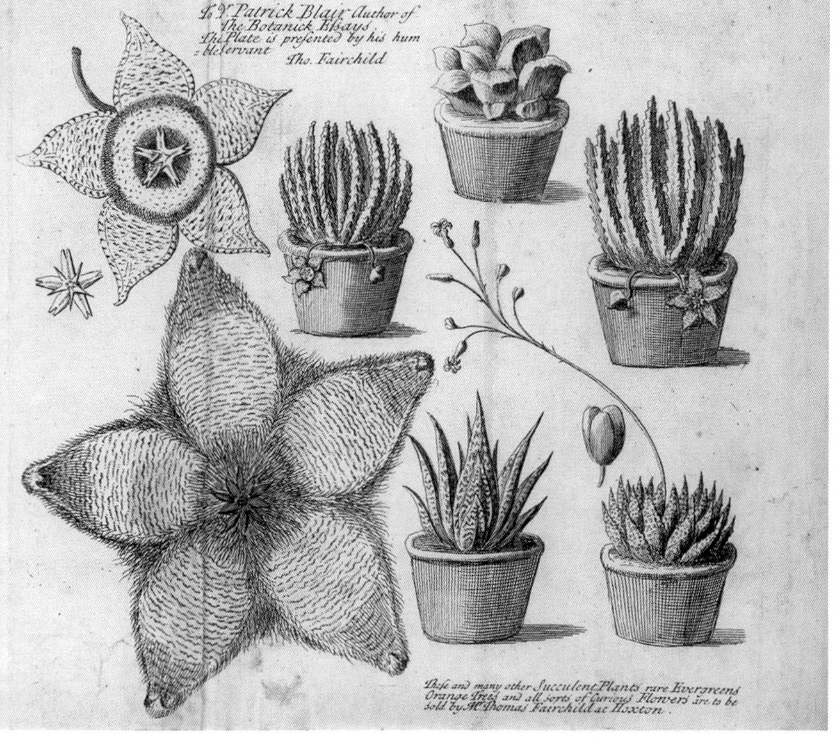

Shoreditch Old Church as Fairchild knew it *(Guildhall Library, Corporation of London)*

Fairchild's tomb off Hackney Road. This stone was carved in 1845 and restored in 1891 *(Nigel Spalding)*

CHAPTER FIVE

The Green-Fingered Fraternity

I dare boldly pronounce it, there is not amongst men a more laborious life than is that of a good gardener's; but because a labour full of tranquillity and satisfaction; natural and instructive, and such as contributes to the most serious contemplation, experience, health and longevity . . . In sum, a condition it is, furnished with the most innocent, laudable and purest of earthly felicities.

JOHN EVELYN, *KALENDARIUM HORTENSE*, 1706

The London Society of Gardeners was formed in the early 1720s at roughly the time that members of the Gardeners' Company were petitioning to have their guild's affairs examined critically. It seems likely that Fairchild and the activists who had signed the petition finally despaired of their interests being properly represented by the company, still bound as it was to its medieval traditions and subject to the machinations of City politics. In any case, the guild had never taken an interest in advancing botanical research. Taken together, those factors would have inspired Fairchild and other curious gardeners to set up an alternative grouping.

The society usually met at the Newhall Coffee House in Chelsea, right on the other side of town from Fairchild and the other Hoxton gardeners who were members. Though inconvenient for them, it was on the doorstep of one of the rising stars of horticulture, Philip Miller, whom Sloane – on Blair's recommendation – had appointed curator of the Chelsea Physic Garden. Little is known of the Newhall: it never achieved the notoriety of Chelsea's most famous coffee house, Don Saltero's at 18 Cheyne Walk, opened in 1695 by James Salter, an exuberant barber and fiddle-player. Sir Hans Sloane, who lived nearby and was sometimes seen at the coffee house, may have helped Salter to set up the museum of curiosities – many of suspect authenticity – that was one of its principal attractions. Perhaps it was to avoid the embarrassment of a possible meeting with his patron Sloane that Miller organised the group in the less known and almost certainly less crowded Newhall.

The son of a Scottish-born market gardener in Deptford, south-east of London, Miller began growing flowers near St George's Fields in Southwark until he was lured by Sloane to Chelsea, first as assistant gardener and then, in 1722 at the age of 31, as the top man. He stayed at the garden until he was dismissed only a few months before his death, aged 80, in 1771. A man of an experimental and enterprising turn of mind, Miller introduced to Britain countless new species sent to him by plant-hunters from all over the world. Under his guidance the Chelsea Physic Garden gained worldwide renown and was visited by Linnaeus in 1734 – apparently a prickly occasion, with the crusty Miller at first sceptical about this Swedish know-all and his newfangled plant names, although he would eventually become convinced of their merit.

Richard Pulteney, in his 1790 *Sketches of the Progress of Botany*, wrote of Miller: 'It is not uncommon to give the term of botanist to any man that can recite by memory the plants of his garden. Mr Miller rose much above this attainment. He added to the knowledge of the theory and practice of gardening that of the structure and characters of plants, and was early and practically versed in the methods of Ray' – whom he had met in his boyhood.

Miller's most lasting contribution to horticultural literature was *The Gardener's and Florist's Dictionary*, first published in 1724 and revised several times afterwards. It lists Fairchild as the only supplier of several rarities, including aconites and the Christmas rose. The book includes endorsement from Miller's fellow members of the Society of Gardeners: 'We whose names are underwritten do approve and recommend this book as highly useful and necessary for all lovers of gardening.' The names – presumably listed in order of importance and prestige – were in effect a roll-call of most of London's most distinguished gardeners and nurserymen of the time:

Thomas Fairchild at Hoxton
Robert Furber at Kensington
Robert Smith at Vauxhall
Samuel Driver at Lambeth
Moses James at Standgate (also in Lambeth)
Obadiah Lowe at Battersea
Christopher Gray at Fulham
Benjamin Whitmill at Hoxton
Francis Hunt at Putney
Wm. Gray Jr. at Fulham.

These men, with Miller, seem to have formed the nucleus of the Society of Gardeners: Henry Wise, a notable omission, may have felt that, with his royal connections, he was too grand to mix with these lowlier toilers. The preponderance of nurserymen from South and South-West London explains why Chelsea was a convenient meeting place. Miller, writing to Sloane in 1725, suggested that the role of the society's members in his dictionary project was not simply to endorse it, but that they had also helped him to write it. At the meetings, members would often bring rare plants and swap information about how they had cultivated them.

They were also concerned with plant classification and identification. In this pre-Linnaean age, the relationship between species was still a confused area, and consequently the naming of plants was often quite random. There was particular uncertainty surrounding fruit, partly because market gardeners, paying no heed to botanical accuracy, tended to invent attractive new names that would encourage customers to try them – an early deployment of marketing techniques in the food industry. It was primarily to bring some order into this complicated field of terminology, so as to curb the rogue fruiterers and plant-sellers, that the Society of Gardeners decided to publish a series of at least four catalogues of its own.

The first of them – and in the event the only one to appear – covered trees and shrubs. Published in 1730 as *Catalogus Plantarum*, it was credited to the above ten nurserymen and ten others, who had presumably joined the society in the intervening six years. Fairchild had died a year before it was published, and the fact that no further volumes appeared suggests that he had a large hand in compiling it. There are some stylistic similarities with *The City Gardener*, but in publishing terms it was a much more ambitious project – a large volume, handsomely illustrated. Fairchild's nephew Stephen Bacon, who took over the Hoxton nursery after his death, is also named as a contributor.

The principal author, whether or not it was Fairchild, possessed an orderly mind. The book's introduction gives a short account of the origins of the society as a club of professional gardeners and explains the problems of nomenclature. It lays the blame for the confusion over plant names on some recent gardening books 'written by persons of slender skill in the knowledge of particular plants, flowers or fruits . . . most of them having contented themselves in copying from other writers, and hence have proceeded many egregious mistakes in borrowing from others the names of plants'. As a result of this, 'gentlemen very often send to nurserymen for the same kinds of fruits, plants, etc. which they already have growing in their gardens, though under a different name, and . . . they presently conclude that the nurseryman who supplied them was either a knave or a blockhead, when in reality the fault was their own'.

Next comes an explanation of what the volume contains and how it is organised. It is 'a catalogue of the several sorts of trees and shrubs which will endure to be planted in the open air in England that are to be found in the several nurseries near London'. Each section deals with a particular genus, whose distinguishing characteristics are described at the beginning. 'Then we have enumerated the different species we have growing in the several nurseries; and afterwards added a short account of their culture.' The plants are listed in alphabetical order according to their then accepted Latin names, from *abies* (fir) to *ulmus* (elm), with an index of English common names that refers back to the Latin. Entries vary in length from about a quarter of a page to two and a half pages for roses, of which forty-three are listed.

The list includes several 'exotics', a name used for any tree that was not a British native. There had apparently been complaints from arboreal nationalists that too many foreign trees were being imported and local ones ignored, but the authors of the *Catalogus Plantarum* are having none of it. Such trees, they declare, are 'by the construction of their parts wonderfully adapted to grow where they can receive but a small share of their nourishment from the earth'. They were 'not only of vastly different natures but also vary as much in their outward appearance, so that by the products of different climates we behold, as it were, a new world'. At the end of the book are 21 coloured plates of botanical drawings illustrating some of the listed trees and shrubs.

The publication was unfortunately timed. Only seven years later Linnaeus made his definitive classification of genera and species, which, when introduced to Britain in the 1760s, made much of the *Catalogus Plantarum* obsolete.

Richard Bradley was obviously a frequent visitor to Hoxton, and a significant amount of the material in his copious writings is gleaned from observations made at Fairchild's nursery and those in its immediate vicinity, in particular Benjamin Whitmill's and Samuel Chapman's. In 1724 Fairchild contributed to Bradley's *General Treatise on Husbandry and Gardening* a month-by-month list of plants flowering in his garden, and an impressive list it is, containing around 70 entries for each spring and summer month but still managing 25 in December. He lists the mule as being in flower in May and July (Philip Miller wrote that one of its virtues was that it flowered twice in a single summer), while 'Fairchild's mule' appears in August. It is unclear whether these references are to two separate flowers. They could be different versions of the carnation/sweet william hybrid, or he may have made unrecorded experiments in crossing other plants.

The fact that he had tulips, carnations and hyacinths flowering from January shows that he was skilled at gardening under glass, and Bradley devotes much space to

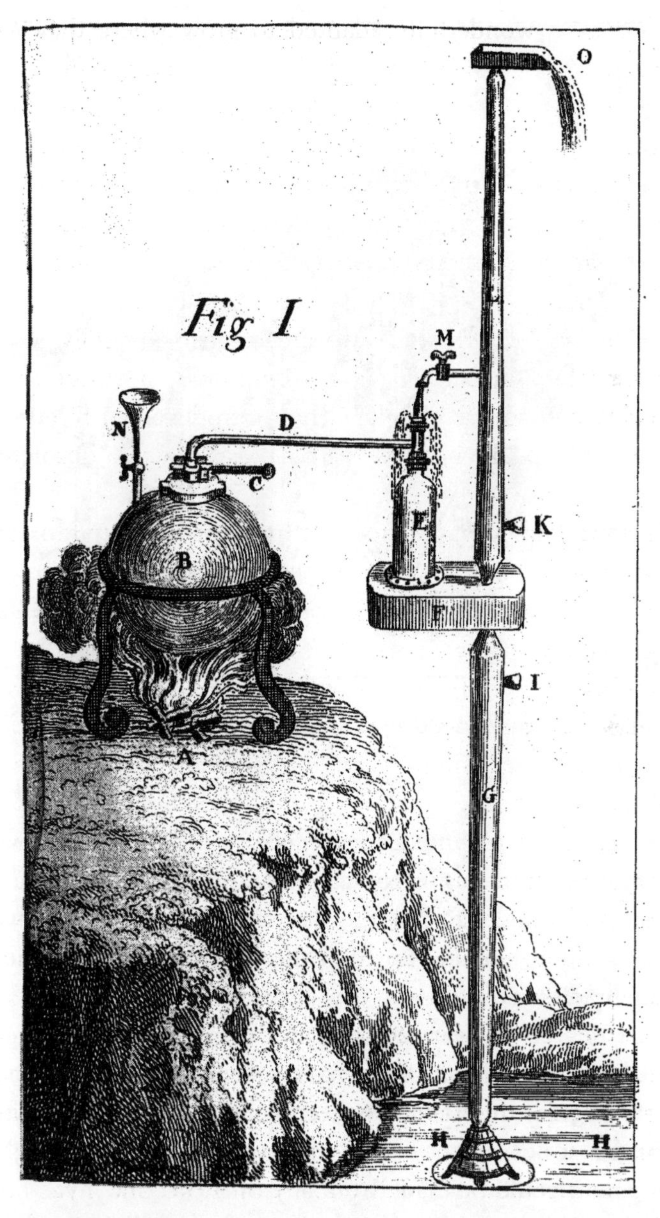

State-of-the-art heating system from Bradley's *New Improvements of Planting and Gardening*, 1726

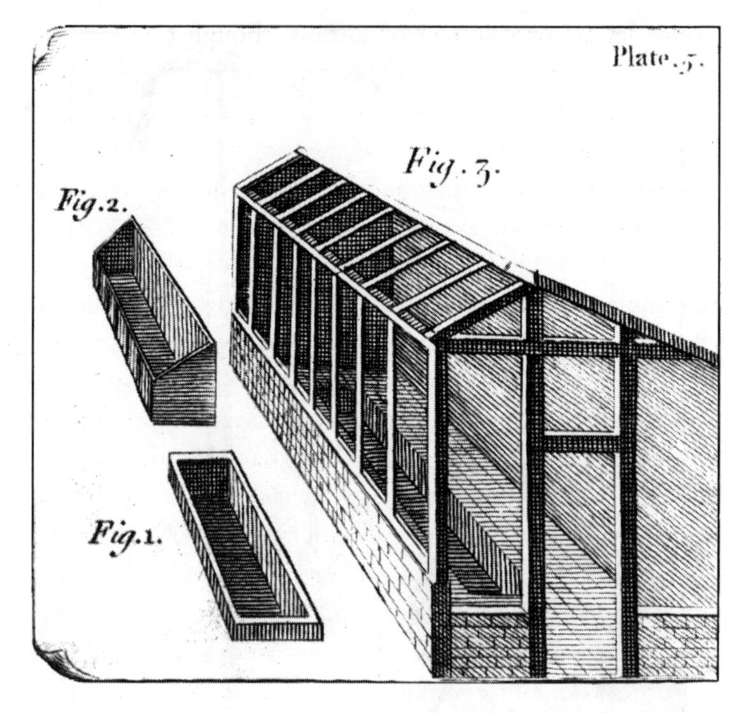

Bradley's design for a stove, or hothouse

describing his techniques, especially a new kind of 'stove' (hothouse) for raising pineapples and the most tender crops. Fairchild had put the flue above the floor, so that any damp air coming up from the ground would be dried out. Bradley pointed out the importance of knowing the exact degree of heat required and of regulating it with a reliable thermometer, 'for I have seen in one place above forty degrees difference between some thermometers with printed scales at the very same time, so that no right judgment could be made from any of them'.

He ends the section on stoves with a friendly advertisement for Fairchild as a baby-sitter for tender plants:

> Nor let anyone despair of success, though they have not stoves immediately of their own to put the plants into which they bring over, so long as there is such a garden as Mr Fairchild's at Hoxton, near London, where such things may be immediately taken care of, and managed with skill.

Fairchild grew many varieties of hyacinth, a flower for which there had been a fad in the late seventeenth century almost on the scale of the earlier tulip craze, with £100 each being paid for bulbs of double varieties. (These had been introduced accidentally by the Dutch grower Pieter Voerhelm, who initially threw away double hyacinths, thinking them an aberration, until he recognised that he might be able to market them successfully.) The monthly lists include 39 varieties of grape, and Bradley makes many references to Fairchild's prowess with this fruit. He says he persuaded the nurseryman to set up a demonstration at Hoxton of how to grow vines. He planted eight or ten varieties and showed how they should be looked after 'for an example to those who are curious to see and observe the manner of vineyard management'.

What is more:

> I observed a sort of vine which had a second crop of grapes, almost ripe about the middle of September, which I suppose might partly happen from an extraordinary pruning Mr Fairchild gave them this year, as well as the extraordinary season.

The summer of 1723 was exceptionally long and warm, but Bradley goes on to speculate: 'Without either of these [the pruning or the fine weather] they would have attempted a double crop, but then without these helps they would not have ripened.' Fairchild managed a second crop of white figs in the same year.

The new method of pruning vines was probably the one described by Bradley in his later book, *New Improvements of Planting and Gardening*. The conventional time to prune them was in winter, leaving four buds on each stem. Fairchild, though, pruned them just after the grapes were gathered, leaving a length of stem proportional to its vigour: about a yard on a strong, plump stem and less on weaker ones. 'By this method of pruning, I observe he never fails of large quantities of grapes, ripening very well, and much earlier than any of the same kind pruned after the usual manner.' Nowadays experts recommend pruning both in autumn and spring.

The article of Bradley's that says most about Fairchild's appetite for experimentation is a catalogue of grafts attempted at Hoxton in 1723:

> For the further improvement of gardening by increasing of plants, even such as will neither grow by cutting or layers, or of such as one cannot readily get any seed of, Mr Fairchild has tried several experiments this and the last year, in grafting by approach or inarching, which are both new and curious. The following is an account of such as have taken and are in prosperous condition.

In all, 21 successful experiments are recorded, starting with 'the terebinthus [turpentine tree] on the pistachio' and ending with 'vines upon vines'. Among the more ambitious is a graft of a cedar of Lebanon on a larch, 'which is the more extraordinary, seeing the cedar is evergreen and the larch drops its leaves'. One flower cross is recorded: a geranium with variegated leaves being bred to a geranium with a scarlet flower.

The rest of Bradley's journal is composed of correspondence from other gardeners and his own wide-ranging observations on horticultural matters. There is an account of a pear tree at Chapman's nursery that bore fruit twice in one summer. Other articles are about garden design, breeding rabbits, growing flowers from seed, mushrooms (again Fairchild's experiences are documented) and transplanting trees.

After nearly 300 years it is hard today to make positive identifications of all the plants in Fairchild's monthly lists, because of the changes in nomenclature introduced by Linnaeus. Yet even if the identities of some of his blooms are still uncertain, the lists show that, given the state of contemporary gardening knowledge, Fairchild developed an enormous range and was always seeking to add to it. Some dried and pressed examples of his flowers have been preserved in the Sloane Herbarium at the Natural History Museum in London and the Sherard Herbarium in the Department of Earth Sciences at Oxford University. Both have examples of Fairchild's mule, dissimilar in several respects.

William Sherard, a cousin of Petiver, was an Oxford graduate and an enthusiastic botanist who travelled extensively as a young man and lived for some years in the Turkish port of Smyrna (now Izmir), where he was made the British consul. On his return in 1717 he began to establish a garden of rarities at a house in Eltham, south-east of London, belonging to his brother James. He collected botanical specimens for most of his life and bequeathed the herbarium to Oxford, along with an endowment to establish the university's chair of botany. The first professor was John Dillenius, who had come to Britain from Germany at Sherard's invitation.

Like many botanists of the time, Sherard sent samples to Petiver, which found their way eventually into the Sloane Herbarium. These included several plants grown from seed supplied by Sherard to the Revd William Stonestreet, rector of St Stephen's Walbrook in the City of London, who gave them to Fairchild to grow, indicating that he already had a high reputation as an experimental nurseryman. Among them are wild flowers such as knotty knawell (knotweed), ash-leaved scabious and yellow oriental knapweed. Other items in the Petiver collection identified as coming from Fairchild's nursery include fleabane, china pink, a lavender-coloured helichrysum (used today as dried flowers), clematis and styrax (aromatic gum).

Such was Fairchild's range and expertise that enthusiasts like Bradley could scarcely keep themselves away from Hoxton. His nursery became part of a circuit that keen gardeners from outside London travelled to on their visits to the capital. In July 1722, for instance, Sherard wrote to his friend and fellow gardener Richard Richardson. After reporting on the selection of Philip Miller as head gardener at the Chelsea Physic Garden, he added: 'Dr Beeston of Ipswich has been in town, at Fairchild's, Chelsea, Hampton Court and Eltham . . . He [Beeston] is very curious and knowing in plants, has a fine collection of exotics.'

Another frequent visitor was James Douglas (1675–1742), a Scottish-born obstetrician, anatomist, personal physician to Queen Caroline and a specialist in gallstones, who achieved his greatest fame in 1726 when he cast doubt on a notorious fraud in which Mary Tofts of Godalming,

Surrey, claimed that she had given birth to rabbits. Douglas was a hobby botanist. In 1719 he interested himself in the germination of mistletoe, apparently with the intention of talking to the Royal Society about it, although there is no record that he actually delivered the paper. In any event he kept a detailed journal of his observations about the parasitic plant, many of them made at Fairchild's nursery.

Thus, on 3 February:

> I had a branch from Mr Fairchild at Hoxton where the flowers began to open pretty wide. They first appeared to him on 30 of last month: perhaps the warm weather which we have had for this 14 days or three weeks did contribute to their early opening.

On 2 March: 'At Mr Fairchild's, the sprouting leaves were much longer shot out and left two smaller ones at their roots.' On 22 March: 'One of the seeds of the viscum [mistletoe] that Mr Fairchild had fixed on the bark of the oak in his own garden, by scraping only a little of the bark down to its middle, was observed to shoot almost a quarter of one inch in length, its having been of a greenish colour . . . This seed was set about April 27. It grew but very little ever since.'

Clearly, then, Fairchild was a meticulous record-keeper. He was also something of a wizard with bulbs, especially those of the Guernsey lily, today known as the *Nerine sarniensis*, that had been a talking point in London gardening circles for several years: in his letters from the Netherlands, Bradley was continually asking Petiver whether Fairchild had one in bloom, and if so could he send it to him.

Douglas was especially fond of the dark pink autumn

flower and confirms that Fairchild was one of its most successful growers in London. That was why the doctor took his precious bulbs to Hoxton and had the master gardener pot them up for him, as he described in a pamphlet he wrote about the Guernsey lily in 1725:

> Whoever will but give themselves the trouble to walk to Hoxton in the months of September or October and view it in Mr Fairchild's garden, in its full prime and beauty, will readily agree that it richly deserves to be taken pains about.

Four years later Douglas revised the pamphlet on the basis of further experiments at Hoxton: 'The ingenious and judicious Mr Fairchild of Hoxton not only communicated to me his own observations concerning the cultivation of this delicate plant but allowed me the free use of all those which grew in his garden, as often as I judged it necessary to examine them.' On the basis of those examinations, Douglas was able to scotch the widely held notion that the root flowered only once in its lifetime: 'Mr Fairchild has often had a root flower again in four years, after the first time.'

He gives the text of a letter Fairchild wrote to him advising on the 'proper management' of the Guernsey lily: 'They love a light earth and a little lime rubbish [rubble] with it does very well: it keeps the roots sound. For if the earth be too stiff or wet, you may keep them many years before they bloom.' He went on to warn that if they are in pots they should be brought indoors for the winter – for pots freeze harder than the open earth and the roots could rot – and then he explains how he manages the roots so that they bloom once every four years.

Like much of Fairchild's horticultural method, this is based on close observation of the way the plants grow. He notices that the flower stalk always comes up in the middle of the root, but the leaves that follow the flower are all on one side of the stalk. This has the effect of changing the central point of the root, leaving the old stalk as part of its circumference. 'There is a new heart made in the middle of the bulb, which is three or four years before it hath strength to blow.' He contrasts this with the tulip, whose leaves feed back into the bulb with enough strength to let it flower every year, though with decreasing vigour.

Problems arise when the lily's leaves are killed off by frost before they have the chance to feed back into the bulb, and when ignorant gardeners cut off the leaves when they are still green, 'which will so much weaken the plants that they may keep them 20 years and not have them blow'. He concludes: 'By the above method of management, where there is a stock there will be continuously some blowing. I hope this account may be acceptable from Sir, your most humble servant, Thomas Fairchild, Hoxton, October 1724.'

Nobody was making better progress than Fairchild in working out the mechanics of plant reproduction and how it could best be manipulated by gardeners who sought to capture for themselves the most exotic creations of the natural world. But he was beset by a nagging doubt. Was he imposing his will too imperiously on the preordained order of things? Was he straying into the rightful preserve of one much mightier than he? Since his life and work are chiefly documented through the writings of others, we can only guess at how or whether he resolved this question of faith.

A large collection of James Douglas's papers is held at Glasgow University, and one other incident from them is worth recording as an indicator of the state of biological knowledge at the time. It also happens to involve the unquenchable Richard Bradley and his talent for grabbing the wrong end of the stick. Douglas had a reputation as an anatomist. In 1720 Sir Hans Sloane, ever anxious to augment the fund of public knowledge, acquired the corpse of an elephant. The beast had been brought alive from India to be displayed to the public but died in 1720 – either from cold, unsuitable food or from being fed ale by mischievous spectators. Sir Hans had the corpse placed on the lawn at his Chelsea house and invited Douglas to dissect it. At first the eminent doctor was reluctant to get involved: 'I have seen the dead elephant, whose enormous bulk quite frightens me,' he wrote to Sloane. 'I am extremely obliged to you for your generous offer but I must desire to be excused for not accepting it, having at present no leisure to manage so great a subject.' So Sloane asked the Revd William Stukeley, a noted physician, to tackle the dissection. In the end Douglas could not bear to think of it proceeding without him and, finding an unexpected window in his schedule, went along anyway. He was given the trunk and sexual organs to dissect and, as the weather was poor, he took them home to study in greater comfort. No discoveries of great significance were made, but the observant Douglas was able to assure Sloane that the elephant was female, not male as had been supposed.

Bradley, when he heard about the dissection, could not resist parading both his own expertise and his unorthodox view on the animal's sex. 'As you have had lately the great opportunity of dissecting an elephant,' he wrote to Douglas, 'I think myself obliged to send you the remarks I made on that creature when it was living, which perhaps might escape your observation.'

When Bradley had visited the elephant the previous August, its tusks were just beginning to appear. His first (incorrect) deduction was that tusks were more akin to horns than to teeth and would grown again if removed. Then he went into a long discussion of the creature's sex and mating habits. He had observed a body part 'somewhat resembling a penis which now and then appeared six or seven inches long, backwards behind the hind legs'. Examination showed that it 'terminated like the penis of a horse'. The same organ was used for the discharge of urine, 'very little at a time but frequently repeated when once this part began to show itself, which used to be commonly in an afternoon'. Then Bradley went into the tricky question of its sex:

> If I remember right, when you dissected the generative parts of this animal you found it to be hermaphrodite, and 'tis a query with me whether all elephants are not so, for, considering the difficulty they must have in coupling, I should not wonder if they were androgynous, as some other animals are which have difficulties in copulation. The unwieldiness of this animal's body seems to make it incapable of acting with the female like other quadrupeds, or even to suffer the female to lie on her back as some authors imagine . . . If what we observed was the penis, which as often as we saw it appeared backwards, so that I rather think they couple by turning their backs to each other, and it may be perform mutually the same offices to one another.

For absolute clarity he attached an explicit little drawing of such an imagined elephantine encounter. Whatever

Bradley speculates on how elephants copulate. 'A' marks the penis (*Glasgow University Library, Department of Special Collections*)

discoveries were being made about sex in plants, sex in animals was still a good deal more riveting – as the producers of television wildlife documentaries would recognise 250 years on.

Customers for Fairchild's trees, vines and lilies would clearly need plenty of space to accommodate them. The concept of town houses crammed together in terraces, with pocket-handkerchief gardens at the back, was only just beginning to take hold. Anyone wealthy enough to own a garden worthy of the name would want it to be big enough to contain several varieties of fruit and, ideally, stately rows of tall trees, keeping up with a fashion that had crossed the Channel from Europe. Travellers returned from France, the Netherlands and Italy breathless with tales of wonderful walks and vistas through avenues of elms, horse chestnuts, limes and poplars. British noblemen, such as the Duke of Beaufort in his palatial seat at Badminton and the Earl of Carlisle at Castle Howard, sought to emulate them and had thousands of trees planted. John Tradescant planted a mile-long avenue of lime trees for the Duke of Buckingham at New Hall in Essex.

Fairchild, as we have seen, did not cater to this bulk market – his nursery would have been nowhere near big enough – so he concentrated on fruit, shrubs, smaller trees and flowers, especially exotics. He may have been fortunate not to have had too many customers among the aristocracy, who were notorious for being slow in paying their bills. In *The Making of the English Middle Class*, Peter Earle tells of a nurseryman who found in 1740 that the only money due to him were two bills contracted in the mid-1730s by Lord Weymouth and the Duke of Marlborough 'which God knows when I shall receive'.

The way for an ambitious nurseryman to gain the attention of potential customers was to offer a mixture of exclusive novelties and reliably productive stock. Keith Thomas, in *Man and the Natural World*, estimates that 89 new species of trees and shrubs were introduced into England in the sixteenth century and 131 in the seventeenth. To keep pace with the novelties, Fairchild throughout his career made a point of learning what the plant-hunters were bringing back from distant lands.

Bradley, on his farcically eventful jaunt to the Netherlands in 1714, was not the first agent commissioned by Fairchild to travel overseas to find material for his nursery. Mark Catesby, later a renowned artist and author of *The Natural History of Carolina, Florida and the Bahama Islands*, visited Virginia between 1712 and 1719 and sent many new plants back to Hoxton, where he may have worked for Fairchild both before and after his trip. This adventure into the New World was plant-hunting in the real, pioneering sense. The Netherlands, after all, had much the same climate as Britain and supported the same kind of plants: it was interesting chiefly because the Dutch were then more advanced in horticulture than the British and because the Leyden Botanic Garden, bristling with exotics from distant Dutch colonies, was a powerful magnet for green-fingered pilgrims from across the water. This was plant-hunting at second hand;

America was much more of a mystery. Who knew what outlandish vegetation might not be found on an apparently limitless new continent thousands of miles across the ocean, with a range of extreme climatic conditions?

Catesby was born in 1682 and spent his childhood in Sudbury, Suffolk. His interest in plants may have been inspired by an uncle, Nicholas Jekyll, who ran an experimental garden and was acquainted with the naturalist John Ray. Samuel Dale, a young associate of Ray – and yet another physician and apothecary who dabbled in botany – lived nearby and befriended Catesby, helping provide the funds for his first journey to Virginia in 1712. There Catesby stayed at Williamsburg, the capital of the colony, with his sister, who had emigrated with her husband William Cocke, also a doctor. Even if Catesby did not work for Fairchild before he set out on the voyage, the two men had certainly met, and he agreed to send plants back to the Hoxton nursery, as well as to Dale, his patron, and to the Chelsea Physic Garden.

Catesby travelled widely and made at least one visit to Jamaica. He sent back hundreds of seeds and scores of plants; whose variety can be judged from Bradley's monthly Hoxton plant lists. Fairchild was meticulous in noting the origin of his plants. One or two he specifically ascribed to Catesby. Others, noted as coming from Virginia, were not all necessarily from Catesby's haul, for the two John Tradescants had earlier introduced a number of Virginian plants, but it seems reasonable to attribute most of those in Fairchild's nursery to Catesby. In the April flowering list were Virginian columbine (aquilegia), sweet molly of Virginia (possibly a form of mallow) and the red-flowered horse chestnut of Virginia: the elder Tradescant had introduced the more common white-flowered variety a century earlier, but the red variation may have been discovered by Catesby.

In May came Virginian astragulus, a member of the pea family, still in flower in June and July. The June list began with the great apocinum of Virginia (dogbane, related to the nerium and periwinkle) and included Virginia yellow jasmine and Virginia purple sunflower (rudbeckia). In July he noted the Virginian martagon (Turk's cap lily), which continued flowering in August. October brought the chrysanthemum tree (a tall chrysanthemum or helenium) from Carolina, and 'Mr Catesby's new Virginian starwort' (*Aster grandiflorus*), followed in November by 'Mr Catesby's fine blue starwort', still in flower in December.

January and February were blank months for transatlantic plants, but March brought the exotically named assarabacca of Virginia, known today as *Asarum virginicum*, a low-growing evergreen with purple flowers. The tulip tree and *Cornus florida*, the spectacular flowering dogwood, were among other American plants that Fairchild introduced to an emerging breed of garden owners hungry for novelty.

As a result of the material he sent back to London, Catesby gained a reputation in botanical circles. Petiver brought some of his discoveries to the attention of the Royal Society in 1715, and Dale had sent a number of his finds to William Sherard. Petiver, Dale and Sherard, along with Sloane and other leading enthusiasts, had been members of the Temple Coffee House Club. Founded in about 1688, the club met on Friday evenings to discuss the latest botanical developments and theories – the intellectual equivalent of the more practical Society of Gardeners. They would certainly have talked about the material that Catesby sent back.

So when the traveller returned to London in 1719, he

found that his fame had preceded him and he had little difficulty in raising funds for a further expedition to America in 1722. Among his sponsors were Sherard, Sloane and Sir Francis Nicholson, governor of South Carolina. Sherard was dismissive of Sloane's efforts to raise money on Catesby's behalf: the two patrons had quarrelled in 1720 over Sloane's refusal to lend Sherard some dried plants in the quantity that he sought. (Sherard retaliated some years later by declining to let Sloane see a collection of dried plants he had been sent from Danzig.)

In a letter to a friend about Catesby's project, Sherard wrote bitterly: 'Sir Hans is ready to promote such designs, wallowing in money: but will not procure a subscription among his friends, as he easily might. I cannot think he has been unsuccessful, having had a large share of all that has come into England, and yet I never had a single plant.' Sherard contrasted this with his own generous instincts: 'I hope he [Catesby] will make me suitable returns, that I may furnish my friends.' For all Sloane's good works, Sherard was not the only man to find him deficient in generosity. Edmund Howard, a Quaker who had dealings with the doctor as an old man, wrote: 'The receiving of money was to Sir Hans Sloane more pleasing than parting with it.'

Because of the sponsorship from Nicholson, Catesby based himself in South Carolina for this second expedition, making trips further south to Georgia, Florida and the Bahamas. He suffered a setback at the start when one of the ships on which he had sent a collection of plants in England in 1723 was plundered by pirates. 'I can expect nothing,' a distraught Sherard wrote to a friend. Later ships did get through, though, and a grateful Sherard agreed to Catesby's request for a loan of £20 so that he could buy a slave to help him with his work.

Some of the plants grown from seeds that Catesby brought back from this expedition are attributed to him in the *Catalogus Plantarum*. They include *Bignonia americana*, the purple trumpet flower; *Fraxinus caroliniana*, the Carolina ash; the kidney bean tree, here termed *Phaseoloides*; and *Barba jovis*, or Jupiter's beard.

As far as Fairchild was concerned, the traffic with Catesby was two-way. In return for the specimens sent him, Fairchild shipped quantities of English plants suitable for growing in the emerging colonies by settlers anxious to preserve something of their garden heritage. We know this because Catesby, before embarking on his second trip to America, left Fairchild instructions on the best way to ensure the survival of the plants on the long westward journey. If Bradley experienced difficulties in having them shipped the few miles across the Channel from the Netherlands, how much harder must it have been to keep them alive on a journey of several weeks across the Atlantic?

'Send them in tubs, not in baskets,' Catesby advised, 'for baskets contribute much to the miscarriage.' It was best to ship them in winter, and October was usually the most favourable month.

> Keep the tubs in the ballast, which keeps them moist and moderately warm . . . for on the quarter-deck they are often wetted with salt water and require the greatest tendance from bad weather, and even with the greatest care they miscarry, as they did with me. It is so hot in the hold in summer that they spend their sap at once, and die, so that it is not a time to send anything.

Fairchild showed this letter to Bradley, who published it in his *Monthly Register of Experiments and Observations*, where he also disclosed a method used by Catesby when he was sending seeds back to England. He would place them in a gourd and seal them up, 'and by that means I have not known them to miscarry, in several parcels which he has sent'. This was a primitive version of the Wardian case, a portable airtight container with windows introduced more than a hundred years later by Nathaniel Ward, which would retain heat and moisture and provide an environment in which tender plants could thrive for long periods.

Those plants that came across the oceans from Asia and the Americas in Catesby's time were, as well, perpetually at risk from shipwreck, for this was a few years before a reliable means of determining longitude at sea had been discovered, so the captains seldom knew exactly where they were. Add to this the occasional hazard of pirates, and it is a wonder that so many plants actually arrived alive.

Catesby was sending back not only plants (live and pressed) and seeds but also sketches of them, and today he is best known for the watercolours of flowers and birds that he undertook initially as a kind of catalogue, informing his sponsors what the plants looked like in their natural environment. He spent more and more time on his paintings and, returning to England in 1726, used them as illustrations for his extensive and impressive *Natural History of Carolina, Florida and the Bahama Islands*. Taking 20 years to complete, this is the book on which Catesby's artistic and literary reputation rests, even if Cromwell Mortimer, Secretary of the Royal Society, who helped edit the manuscript for its 1748 publication, went a little over the top when he called it 'the most magnificent work I know of since the art of printing has been discovered'. Linnaeus used some of the illustrations as source material for his definitive classification of plants.

George III bought many of the original paintings in 1768, and they are still in the royal collection at Windsor Castle.

After his return to London in 1726, Catesby went to live in Hoxton, where he worked for Fairchild. It is possible that he took the place of Thomas Knowlton, who would become one of the most celebrated gardeners of the eighteenth century. Knowlton, whose first job was at Sherard's garden in Eltham, had been working for Fairchild for a few years, and in 1726 went to the Netherlands to seek rarities on his employer's behalf. (In 1728, Knowlton became head gardener to the third Earl of Burlington at Lanesborough in Yorkshire, where he stayed until his death in 1782.) When Fairchild died in 1729, Catesby continued to work for Stephen Bacon and stayed at Hoxton until 1733, when he moved to Christopher Gray's nursery in Fulham.

To rely for his stock in trade on chance discoveries by freelance plant-hunters was of no real satisfaction to a man whose attitude to life and business we would today describe as proactive. It was all too haphazard; anathema to a methodical plantsman. Fairchild could grow things that others could not because he took pains to understand the physiology of plants and why each one needed particular treatment and conditions to thrive. Part of his study of how they lived and developed involved examining how they reproduced. Until his time, scholars who had broached the question of sexuality in plants – Grew, Ray, Camerarius – had been botanists rather than gardeners. Fairchild, the observant gardener, was in a better position than the academic botanists to see how the processes they had worked out theoretically operated in practice from day to day, and to discover ways by which human intervention could affect the outcome.

From the start, as Bradley recorded, Fairchild made himself expert in the techniques of grafting, of joining a cutting from a good-tasting variety of fruit on to a root-stock, sometimes of a more compact size, that would ensure it grew into a healthy tree. This is what allows us today to grow fruit even in quite small gardens. He was among the few to cultivate grapes in London successfully, and as a result of a detailed study of how trees form branches he popularised the training of fruit on espaliers. A variety of nectarine that he introduced, Fairchild's Early Nutmeg, still appeared in nursery catalogues nearly a hundred years after his death.

It was his attempt to understand what made a successful graft that drew him to investigate the movement of sap in trees, and from there into the wider topic of the reproductive systems of plants in general. Today, we know that sap in trees rises through the roots and the trunk in spring, producing leaves and flowers and new branches, before descending to the roots again, this time flowing just underneath the bark. That is why grafting fruit trees works. If the join is made properly the sap will flow up from the root system of the host tree – usually chosen for its vigour or growth characteristics – to the newly attached stem of the other tree, selected for the quality of its fruit.

In Fairchild's time there was still much discussion of the mechanics of sap circulation, and he performed many experiments to find the answer. Once again, it was his friend and patron Patrick Blair who conveyed the results of Fairchild's early experiments to a wider public. In his 1720 book of essays, Blair described how Fairchild had practised what was then called circumcision on a pear tree trained against a wall, giving a fuller account of the process than he had in the Royal Society paper in which he also referred to Fairchild's mule.

A ring of bark, some three inches wide, had been cut away from each of the tree's three principal branches in May, after the sap had risen. In September the bark above the rings became swollen with the accumulated sap, unable to descend further. The following season the tree produced an abundance of buds, leaves, flowers and fruit above the incision, but no new shoots. This continued for several seasons until the bark healed on two of the three branches, allowing the sap to flow right down to the roots. Henceforth the tree produced less fruit but began to grow shoots again. The third branch, where the bark remained separated, came into flower earlier than the other two (presumably because the sap had less far to travel) and continued to fruit plentifully.

In time the stem of the tree above the cut grew much plumper than the stem from the incision down to the roots. Blair explained: 'This augmentation in the bigness of the branch clearly demonstrates how the sap, being interrupted in its descent, immediately returns towards the top; and that the circulation is as well maintained from the incised part as from the root.' It was important to leave some branches uncircumcised, so that their sap could return to the roots for their nourishment. Blair compared the circulation of sap to the circulation of blood in animals, likening the tubes through which the sap flowed upwards to human arteries, and the return route to the veins: 'Thus in trees the bark is analogous to the skin in animals, the wood to the bones and the pith to the marrow.'

The dividing line between the study of plants and the study of animal life – including man – was a fuzzy one.

Bradley published theories on both. In 1721 he produced a pamphlet entitled 'A Philosophical Account of the Works of Nature . . . to which is added an Account of the State of Gardening in Great Britain and other Parts of Europe'. In it he sought to divide mankind into five classifications, rather like the genera of plants:

> We find five sorts of men: the white men, which are Europeans, that have beards; and a sort of white man in America (as I am told) that only differ from us in having no beards. The third sort are the malatoes [mulattos], which have their skins almost of a copper colour, small eyes and straight black hair. The fourth kind are the blacks, which have straight black hair, and the fifth are the blacks of Guinea, whose hair is curled like the wool of a sheep, which difference is enough to show us their distinction.

He went on to express what we today regard as an enlightened view in the nature v. nurture controversies: 'As to their knowledge, I suppose there would not be any great difference, if it was possible they could all be born of the same parents and have the same education, they would vary no more in understanding than children of the same house.'

Despite his constant attempts to ingratiate himself with the aristocracy, Bradley seems to have shared some of Fairchild's advanced social views. In the same pamphlet he recommends that uncultivated common land be divided up among poor people for cultivation – a precursor of the allotment system. In this way the poor, 'which at present are troublesome and expensive', may be engaged in productive labour and 'live in a contented state'.

In relation to plants, Bradley became convinced, like Fairchild and Blair, that circulation of the sap was the key to growth. He discussed this at length at the very beginning of his *New Improvements of Planting and Gardening*. 'Life,' he declared, 'whether it be vegetable or animal, must be maintained by a due circulation and distribution in the bodies they are to support . . . The sap circulates in the vessels of plants much after the same manner as the blood doth in the bodies of animals.'

His version was that the juices that sustain life are sucked into the roots from the earth. As the roots are warmed, the juices travel upwards in the plant as steam, triggering the growth of buds, leaves and flowers. When the steam reaches the top of the plant and begins to cool, it turns back into liquid and slowly descends to the roots again. This theory was clearly influenced by the experiments in steam power which had been made since the end of the seventeenth century. Later in the book Bradley described a crude form of steam engine designed by Thomas Savery, used to pump water in the greenhouses at Robert Balle's Kensington garden, and likened its function to the motion of the sap.

Much of the evidence he cited for his theories was gleaned from Fairchild's experiments. A plain-leaved jasmine had been 'envenom'd' by a striped one (an interesting use of the word, implying that the stripe was seen as being produced by a venom or poison in the sap) and produced striped, i.e. variegated, leaves. A bed of lilies in Fairchild's nursery had been struck by lightning one summer 'in the height of their sap' as they were about to come into flower. The following year scarcely one in a hundred flowered, and even fewer the year after that. Bradley added mysteriously that Fairchild had made other experiments that proved the circulation of sap 'but I have not his leave yet to mention them'.

Bradley had a short way with people who disagreed with him. In his *General Treatise* in May 1722 he wrote 'An Answer to some Objections lately made against the Circulation of the Sap', describing his critics as 'not of the first rank among the learned'. (Among them was the gardener and seedsman Stephen Switzer.) He continued caustically: 'I confess I am not in any way displeas'd at any objections that may be made by such people against my writings, as it gives me opportunity of setting them to rights, which I shall always be ready to do.'

Such belligerence, along with his lack of probity, would help to account for his evident unpopularity. Writing of an experiment where Fairchild budded a passion flower with spotted leaves on to another variety with long fruit, he noted that two weeks after the graft the plant had begun to produce spotted leaves. This seemed to Bradley to justify his views on the movement of sap, and he poured scorn on those old fogies who still contested them:

> I am not insensible that when I write my works fall into the hands of two sorts of people; the one who, desiring to be informed, are curious and inquisitive, and would willingly learn; and the other who, finding themselves men by the number of their years, are either ashamed of asking questions lest they should seem ignorant, or else think that their age is sufficient warrant for their obstinacy and talking of nonsense. For the first, I have that charity and generosity that I shall always, as far as my time will permit, think myself well employed in instructing them; but for the latter who are sure they know enough already, and

> resolve against improvement, they are only fit to accompany one another.

Such men as Blair, Bradley and Douglas served to connect Fairchild to the world of science that would otherwise have been closed to him through his lack of formal higher education. Blair and Douglas were both physicians by profession but were among that group of amateur scientists, with Sloane its acknowledged leader, whose intense curiosity drove them to dabble in the rapidly evolving world of botany. They needed someone knowledgeable and competent to carry out the practical experiments by which they could test their theories, and Fairchild fitted the bill perfectly. Their writings bristle with compliments to him.

Blair, born in Dundee, had trained as a doctor and first gained celebrity in London in 1713, when he dissected the bones of an elephant (clearly an animal of immense interest to the scientists of the time) and gave the Royal Society a discourse on the experiment. That same year he became a fellow of the Society, sponsored by Petiver. The *Botanick Essays* were among his most substantial works, and in the preface he gave generous credit to Fairchild for having carried out many of the experiments on which his conclusions were based:

> The lateral tendency of the sap when interrupted in its ascent, analogous to that of the blood at an amputation, is obvious from an experiment of Mr Fairchild's (whom I have often mentioned and to whom I owe all the practical observations I have advanced concerning this vegetation). He cut the stalk of a white lily from the root and topped it when it began to flower, and in

> a short time it pushed forth bulbs from the sides of the stalk, which when put into the ground sent forth fibres and became roots.
>
> He observes that if a tree is planted in the autumn it ought not to be topped until the spring following, for the sap circulates more agreeably when allowed to ascend directly to the top of the autumnal shrub than when interrupted by cutting it off at the planting.

Fairchild makes reference to this several times in his writings, and spring pruning is still the recommended method for most trees and shrubs.

Blair presented several papers to the Royal Society – apart from the one about Fairchild's hybrid – which took botanical knowledge significantly further. He worked with other gardeners as well as Fairchild. On the last day of 1721 he wrote a letter to Sir Hans Sloane – published in the Royal Society's *Philosophical Transactions* – about some experiments carried out by Philip Miller a year before he became head gardener at Chelsea Physic Garden. In one, he separated male spinach plants from female, which went on to produce barren seed. The same thing happened when he removed all the male flowers from a melon.

The third observation was a result of an accident rather than an experiment. Miller had bought some cabbage seeds from a neighbour and found that they came up a mixture of red and white cabbages and savoys, with some containing elements of all three. He discovered that the neighbour had planted one row of each kind quite close together, resulting in what we today call cross-pollination. Blair saw this as a conclusive argument against Ray's theory (advanced at the time by Bradley, among others) that every speck of pollen contained a miniature version of an actual plant, and variations in colouring were the result of weakness in the sap caused by poor planting conditions. 'For if the individual plant be in each grain of the male farina, how can it be so far dismembered as that one part shall go to the making up of the ribs of red cabbage and another to compose the rest of a Savoy plant?'

He went on to speculate that this could be the cause of the 'infinite variegations and stripes' seen in both annual and perennial flowers. 'Were I to extend this to a great many other plants, and were there proper observations made upon them, considerable improvements might be made on the doctrine of the sexes of plants.' He did not always rely on professionals to conduct his experiments for him, and was not averse to trying things for himself. In 1721 he told the Royal Society that he had obtained new varieties of peas by cutting the anthera off some plants and inserting the pollen of another variety into the blossom.

That Fairchild gained the friendship and confidence of both him and Bradley says much for his powers of diplomacy, for Blair was clearly one of the many scholars who distrusted Bradley and took every opportunity of denouncing his theories in his letters and papers. Bradley's response to the letter to Sloane was, for him, rather moderate. In a pamphlet in 1724 he wrote: 'Even Dr Blair allows that plants have a mode of generating, but does not agree with me about the manner in which it is performed.'

Blair was, quite literally, lucky to be alive to pursue his researches and to tell the Royal Society about Fairchild's hybrid. A Jacobite, agitating for the return of the Stuart monarchy, he was arrested at the time of the 1715 uprising that followed the accession of George I. He admitted that he acted as physician to the Scottish rebels, although he claimed in a letter to Sloane that he was in effect press-ganged into it – he never carried arms, he protested, and was never paid. He was sentenced to death none the less and held in Newgate Prison awaiting his fate. His medical and botanical friends, led by Sloane, began pulling strings at court to secure his pardon, but their success remained in doubt until the very last moment.

Petiver organised the effort – indeed, his natural role in life appeared to be as a coordinator and conduit for messages between members of this informal guild of curious botanists. In a letter to him from prison, Blair complained that Sloane, the member of their circle with most influence at court, seemed not quite to appreciate the urgency of the situation. By the eve of the date fixed for the execution, nothing had been resolved. Many of Blair's friends, including Petiver but not Sloane, went to Newgate to sit with him during the anxious hours that could have been his last. Petiver later wrote to Sloane, giving a dramatic account of the nerve-wrenching evening: 'The doctor sat pretty quietly till the clock struck nine, and then he got up and walked about the room. At ten he quickened his pace, and at twelve, no reprieve coming, he cried out: "By my troth, this is carrying the jest too far."' The reprieve finally came shortly before dawn, and Blair eventually won an official pardon.

CHAPTER SIX

Playing God

If you look into our gardens, annexed to our houses, how wonderfully is their beauty increased . . . Our gardeners moderate [nature's] course in things as if they were her superior.
WILLIAM HARRISON, DEAN OF WINDSOR, *DESCRIPTION OF ENGLAND*, 1577

Seventeenth-century scientists and philosophers wrote long and often tortuous tracts about how the world and its wonders were created, invariably concluding that a divine master plan was the only possible explanation. The intellectual and theological climate of the time would allow no theory that did not embrace the orthodox belief in the omnipotence of the Almighty. As scientists discovered truth after truth that seemed to militate against such a belief, the more complicated their explanations had to become.

Even the clear-sighted Isaac Newton felt that he must find a place for God's scheme of things in the arrangement of the world whose mysteries he was otherwise helping to explain so rationally. In his 1997 biography *Isaac Newton: The Last Sorcerer*, Michael White describes Newton's philosophy, as it applied to the laws of gravity:

> God does not himself control directly the gravitational forces that keep the planets in motion, nor does he provide directly the medium via which universal gravitation operates. Instead, the incorporeal ether which facilitates the phenomenon of gravitation (and

> perhaps other forces) is actually the body or spiritual form of Jesus Christ.

And more than a hundred years later, Charles Darwin was so tormented by the religious implications of his work on evolution that he delayed releasing it until another scientist began to publish articles that he feared would pre-empt his discoveries.

The naturalist John Ray – who died in 1705, the year in which Newton was knighted – was a devout churchman as well as a pioneer of English botany. In addition to his learned books on plants, he combined his passions for natural history and theology in 1691 to write *The Wisdom of God Manifested in the Works of Creation*. In it he asked:

> Why should some plants rise up to a great height, others creep upon the ground, which may perhaps have equal seeds, nay, the lesser plant many times the greater seed? Why should each particular so observe its kind, as constantly to produce the same leaf for consistency, figure, division and edging; and that though you translate it into a soil which naturally puts forth no such kind of plant? . . . What account can be given of the determination of the growth and magnitude of plants from mechanical principles, of matter moved without the presidency and guidance of some superior agent?

Ray went on to list the various attributes of plants and their functions, concluding that there had to be 'some intelligent plastic nature, which may understand and regulate the whole economy of the plant'. That was confirmed, in his mind, by the ingenious methods used by plants to produce seed to ensure the continuation of their species. Some seeds, for instance, have feathery attachments that allow them to be disseminated far and wide by the wind. He embraced the view of the structure of seeds that prevailed in the late seventeenth century, and that fits best with the notion of God's forward planning:

> Most seeds having in them a seminal plant perfectly formed, as the young is in the womb of animals, the elegant complication thereof in some species is a very pleasant and admirable spectacle; so that no man that hath a soul in him can imagine or believe it was so formed and folded up without wisdom and providence.

In Ray's view, the weight of the evidence confirmed beyond any reasonable doubt that the vegetable kingdom was a divine creation. One final observation seemed to clinch it. He noted that in countries where particular diseases were rife, there was invariably an abundance of plants that could treat them: 'So in Denmark, Friezland and Holland, where the scurvy usually reigns, the proper remedy thereof, scurvy-grass, doth plentifully grow.' His contemporary, Nehemiah Grew, although one of the first to be convinced of the sexual nature of plants, shared Ray's faith in divine omnipotence. In his last work, *Cosmologica Sacra, or a Discourse of the Universe as it is the Creature and Kingdom of God*, published in 1701, he made much the same arguments as Ray.

Keith Thomas remarked in *Man and the Natural World*: 'It was in the later seventeenth and early eighteenth century that these arguments about the Creator's design reached their most ingenious and fanciful.' It was, in effect, compulsory for writers on gardening and botany to pay a ritual compliment to the omnipotence of the Creator. The last chapter of Robert Sharrock's *History of the Propagation and Improvement of Vegetables*, published in 1660, was entitled 'The conclusion of the treatise, with one or two choice observations of the wise and good providence of God, which may be seen in the admirable make of vegetables, and fitness to their ends.' In 1682 Samuel Gilbert wrote in *The Florist's Vademecum* of 'each flower showing the providence of the almighty God, and that we may read Him in these His beautiful handiworks that so diaper our gardens'.

In the mid-eighteenth century Charles Alston, the King's Botanist in Scotland, wrote that the realisation that all living things were of God's creation would 'render distasteful to all, wanton and useless destruction of God's works and especially of life, even amongst lower manifestations, for mere idleness or sport' – an argument very close to that of present-day environmentalists. He concluded: 'So the perfection of reason harmonizes with the perfection of faith. God's two voices coincide.'

The underlying assumption was that God had designed the world and furnished it with useful plants and animals specifically for the wellbeing of His most important creation, mankind. Some scholars believed that every plant was designed to be useful to mankind in some way and that the shape and colour of its leaves and flowers constituted a divine code that indicated its particular use: for instance, a yellow flower might cure jaundice, and one with spotted leaves get rid of facial spots. Most accepted that God had created a fixed number of genera and species, and that they had remained constant since the Creation – a view that was held even by the revolutionary Linnaeus.

It cannot be proved that Fairchild read any books on this subject – except those of Bradley's to which he subscribed

– but it seems probable that, as a serious man about to get involved professionally in gardening, he would at least have made himself familiar with Ray's work, first published when Fairchild was 14. That he took his faith seriously is apparent from his will, where he left money for a sermon to be preached annually in St Leonard's Church in Shoreditch on a choice of two topics, one of which may have been inspired by the title of Ray's book. The lecture, still given today, is on 'The Wonderful World of God in the Creation' or 'The Certainty of the Resurrection of the Dead proved by the Certain Changes of the Animal and Vegetable Plants of the Creation'.

A devout man, he was determined that nothing should interfere with that certainty. Less than 50 years after James II had been deposed for his Roman Catholic predilection, it was still a time when a man was largely defined by his religious beliefs and the outward show he made of them. There were many dissenters from the teachings and practice of the established Church of England – on the one hand the traditionalist Catholics who remained loyal to the Pope, and on the other the radical Puritans who believed that the Anglican Church was still riddled with Roman practices – but they were still regarded to some extent as outcasts. After all, had not Fairchild's friend and patron, Patrick Blair, nearly lost his life through supporting a rebellion that had its roots in religious differences?

Yet at the same time Fairchild, the hard-headed business-man, was driven by an equally powerful commercial imperative. In the early years of gardening, competition in nurseries had been based almost entirely on patronage, and to a large extent it still was. Whom you knew was critical. That was how London and Wise, who had worked as gardeners for the aristocracy, had built up such an impressive customer base for their Brompton Park nursery among the owners of the great estates. Fairchild, who is not known to have worked as a gardener at a big house, made a virtue of being a man of the people and appears to have enjoyed little top-drawer patronage. Thus he was forced to compete through the excellence of his techniques and the breadth of his stock. That meant the provision of new varieties – and one means towards that was to undertake the experiments that would eventually bring him dangerously close to conflict with his faith.

Like today's opponents of genetic engineering, Fairchild was plagued with doubts about the morality and the possible long-term effects of what he was doing. His dilemma was that his own observations, and the experiments he conducted based on them, could by some constructions be seen to contradict the spirit and the letter of his religious beliefs. A few years later, Linnaeus would face the same difficulty with the Roman Catholic church, which banned his books on plant classification and ordered them to be burned. The Age of Reason may have been about to dawn in many parts of Europe, but it would be a while yet before its light began to break over Rome.

Fairchild's bequest for the Shoreditch sermon was a way of expiating his guilt. It would allow men far more pious and learned than him to argue against what his experiments appeared to prove: that it was possible to override the divine will and create living matter of man's own devising.

The old church of St Leonard's in Shoreditch, where Fairchild worshipped, occupied the same site as today's Georgian church, at the busy junction of Hackney Road, Kingsland Road, Old Street and Church End (now

Shoreditch High Street). Although the streets around were lined with tenements and other buildings, much as they are today, just beyond them were open fields and orchards. A little south were the ruins of the medieval Holywell Priory and the Curtain Theatre, one of the earliest in London. A few hundred yards north, along Kingsland Road, the Geffrye Almshouses – now the Geffrye Museum – were erected in 1715.

The exact date of the church's construction is unknown but, like most medieval churches, it was enlarged over the centuries from modest beginnings. Drawings from Fairchild's time show a squat building with a square tower at the western end surmounted by a hexagonal bell-cote, domed at the top. The top section of the tower was clad in weatherboarding. There were four aisles, each with a large, ornate Gothic window at the east end. Most churches have no more than three aisles: the most northerly of these four is believed to have been added as a chantry chapel in 1482. Between 1581 and 1630, galleries were built on the west, north and south sides of the nave to accommodate the rapidly expanding population of the parish. Among the memorials was one to Cuthbert Burbage, who managed the Globe Theatre and whose brother, Richard Burbage, acted in and produced many of Shakespeare's plays.

Towards the end of the seventeenth century the church began to show its age, and from 1675 onwards a series of repairs were carried out. They served to slow the building's decline but not to halt it: the fact that the floor of the nave was below the level of the surrounding land made it vulnerable to decay. In 1700 the chancel was raised, which improved matters but also provided an opportunity for thieves to hide in the church and steal the gold and silver embroidery from the pulpit cloth.

On 4 April 1711 the parishioners petitioned Parliament to have the whole church rebuilt:

> Setting forth the extreme poverty of the petitioners and the ruinous condition of their church, the same being at least six feet under ground, and many departing and absenting from public worship for want of a place in the said church; all which may be remedied by the rebuilding of the old parish church and building a new church, according to the scheme now lying before your honours, and praying that out of the money granted towards building of churches, such a sum may be granted as be thought fit to rebuild the said church.

The petition failed. About two years later part of the bell-cote was blown off in a gale, and three years after that, on 23 December 1716, the parishioners' worst forebodings were fulfilled. According to a contemporary report:

> . . . the walls of the old church rent asunder with a frightful sound during divine service in the afternoon, a portion of the tower also gave way and a considerable quantity of mortar falling, the congregation fled on all sides to the doors, where they severely injured each other in their efforts to escape.

Twenty years later the old church was pulled down and the new one erected to the design of George Dance the Elder.

There is no saying whether Fairchild was present at that alarming event two days before Christmas in 1716. But we do know that he was a rational and modest man. Had he been at all superstitious and self-important, he might have been tempted to see the collapse of part of the church as an expression of God's anger at the interference with nature's grand design that had been perpetrated at the nursery a few hundred yards away – an impertinence that was about to be revealed to the world.

Bradley's 1717 description of Fairchild's mule is the first record of it, but the original cross was probably made some years earlier. Carnations and sweet williams, among the oldest and most colourful flowers of summer, would have been part of the stock in trade of any nursery, as they are of the modern garden centre. Both belong to the genus *Dianthus*, which means 'divine flower' in Greek, suggesting that the ancient Greeks valued them highly. Carnations were introduced to England from southern Europe soon after the Norman Conquest by monks who used the clove-scented flowers to flavour their wine. They thrived near lime rubble and were therefore ideal for growing near abbey walls. Originally known as clove-gillyflowers – Chaucer refers to them as such in Sir Topaz's prologue in *The Canterbury Tales* – they had single flowers and came in three principal colours: white, pink and crimson.

The name 'carnation' – probably deriving from 'coronation', because the flowers resemble a crown – began to be used in the sixteenth century for the showier double varieties that were brought to England from the Netherlands. The botanical name is *Dianthus caryophyllus*. An enthusiasm developed for flowers with striped, flaked and speckled petals – the product of natural, random hybridisation or viral infection, to which the species seems especially prone. In *The Winter's Tale* (Act IV, Scene III), written in 1611, Shakespeare includes a passage about striped gillyflowers which tantalisingly suggests that he may have had an inkling about how plants reproduce. Perdita says to Polixenes:

> The fairest flowers o' the season
> Are our carnations and streaked gillyvors,
> Which some call nature's bastards; of that kind
> Our rustic garden's barren, and I care not
> To get slips of them.

Polixenes asks:

> Wherefore, gentle maiden, do you neglect them?

She replies:

> For I have heard it said
> There is an art which, in their piedness, shares
> With great creating nature.

In other words, she has heard that men may have interfered with the flowers to give the striped effect. In response, Polixenes sticks to the orthodoxy that nature – meaning God – is all powerful:

> Say there be;
> Yet nature is made better by no mean,
> But nature makes that mean; so, o'er that art
> Which you say adds to nature, is an art
> That nature makes.

He goes on to describe the process of combining the desirable qualities of plants and trees through grafting – a technique that does not amount to hybridisation but has the same objective:

> You see, sweet maid, we marry
> A gentler scion to the wildest stock,
> And make conceive a bark of baser kind
> By bud of nobler race. This is an art
> Which does mend nature – change it rather; but
> The art itself is nature.

And he exhorts Perdita:

> Then make your garden rich in gillyvors
> And do not call them bastards.

In his book *Old Carnations and Pinks*, C. Oscar Moreton says of the Elizabethan period: 'No farmhouse garden was considered complete without its carnations.' They were the most fashionable of all flowers until the 1620s, when the tulip craze robbed them of that status. But they remained popular, and when Ray wrote his *Flora* in 1676 he was able to list 360 varieties. Just 50 years later Bradley estimated that there were close to 1,000.

The sweet William is *Dianthus barbatus*. The origin of the popular name is uncertain: some say it refers to William the Conqueror, others to St William of Rochester (1154–1226) and others that it comes from the French 'oeillet', a little eye. (Suggested attributions to William III or William, Duke of Cumberland, cannot be sustained, since the name is first recorded in the sixteenth century.) The sweet William is distinguished from the carnation – and from the pink, its more delicate sister – by having a mass of flowers packed on to a single head.

To breed reliable plants for sale, in predictable colours and forms, seventeenth-century nurserymen used cuttings from existing plants. Although some still believed that a seed contained a miniaturised version of a whole plant, and that no external interference was needed to make it grow, they observed that seeds did not always breed true to their known parent: seed from a white carnation might produce pink or striped flowers.

When botanists began seeking explanations for this, they came to realise slowly that other plants had to be involved in the process by some kind of sexual activity that did not involve physical contact between the parents. The invention of the microscope in the 1670s, by the Dutch scientist Antone van Leeuwenhoek, allowed them to identify the organs through which fertilisation might occur. Ray and Grew were the first British botanists to understand the importance of the discovery.

To produce seed, pollen has to be transferred from a flower's male organ – the stamen – to the pistil, the female organ that contains the ovules. The pollen is first deposited on the stigma, the surface of the pistil, where it germinates and makes its way to the ovules at the pistil's base. There it produces the seed. Although nearly all flowers are bisexual, incorporating both stamen and pistil, they are rarely able to fertilise themselves without the assistance of some external agent such as insects (especially bees), wind and occasionally birds. Pollination by insects was only just being mooted in the early eighteenth century, and Bradley was one of the first to record it.

The process of pollination is a fairly random one, although viable seed will not generally be produced from the transfer of pollen between different species. Some plants, too, have built-in defences against self-pollination: the stamen may produce its pollen before the pistil is sufficiently developed to cope with it. This prevents inbreeding which, as in animals, can result in weakening the stock. The parallel with animals is fairly close: not many can breed between species, and in most species there are barriers in culture or instinct against breeding with close relatives.

Because carnations and sweet williams belong to the same genus, there must have been many other spontaneous crosses between them before the one that first caught Fairchild's eye. Most nurserymen would have discarded them as being rogue flowers (or bastards, as Shakespeare so accurately described the striped gillyflowers). But Fairchild had already shown interest in the way plants reproduce: he had discovered, for instance, that if you remove the gland that contains the nectar, flowers will not produce seed. Having found a specimen that displayed the attributes of both a sweet william and a carnation, it would have been hard for a man with such an enquiring mind to resist trying to replicate it.

According to Bradley, Fairchild took the farina (pollen) of a sweet william and placed it on the pistil of a carnation, probably using a feather or a small brush, or perhaps the tip of a quill pen. He would presumably have performed the experiment with a number of pairs of flowers and would have had to keep them isolated to prevent any unplanned fertilisation. Then he would have gathered the seed and labelled it, for sowing the following spring.

By July, as he gazed at the unusual flower, he would have known that he had successfully produced a hybrid, though whether he was aware of the historic nature of his feat is doubtful. He does not mention it in *The City Gardener* and there is no record that he ever claimed credit for it himself, leaving Blair and Bradley to blow his trumpet. He obviously thought the experiment worthwhile, though, since he kept mules of varying descriptions in his nursery stock for some years, and the plant is thought to have been grown until late in the eighteenth century.

Not all the hybrids would have looked the same. Like children of the same parents, some would display more of their carnation heritage, while others would be closer to the sweet william. Of the two dried specimens to have survived, that in the Sherard Herbarium at Oxford is hard to distinguish from a carnation, but the one in the Sloane Herbarium at the Natural History Museum in London has a group of small flowers as its head, like a sweet william. Peter Collinson, the naturalist, wrote to a friend in 1740 that he had a mule in his garden descended from one of Fairchild's original crosses. It had slender leaves and the double flowers of the carnation but they appeared in clusters, as on the sweet william.

At about the time Fairchild was making his initial hybrids, an experiment was being undertaken thousands of miles away, in Boston, New England, that helped build the case for sexual reproduction in plants. Cotton Mather, a Puritan best known for his role in the Salem witchcraft trials of 1692, reported two significant incidents in a letter to Petiver in July 1716. In the first, a row of coloured corn was sown in a plantation of the common yellow corn. Several rows of the yellow corn, those closest to the coloured variety, contained many coloured cobs, the majority of them in the direction of the prevailing wind. The other experiment was in planting inedible gourds among edible squashes, when the bitter taste of the gourds affected the squashes and made them inedible.

Bradley had been doing some experimenting, too, probably in Robert Balle's garden in Kensington. In *New Improvements* he described how he removed the apices, or stamens, from some tulips before they had produced pollen. Planted as they were some distance from any other tulips, they failed to produce seed. He also tried it with auriculas: 'The yellow and black auriculas, which were the first we had in England, coupling with one another, produced seed which gave us other varieties, which again mixing their qualities in like manner has afforded us little by little the numberless variations which we see at this day in every curious flower garden.' To prove this, he isolated a particular variety of auricula in a remote part of the garden and found that its seed bred true; but seed from flowers growing in mixed beds produced many variations in colour.

For all his weaknesses of character, and his enthusiasm for the freak-show school of horticultural experiment, the case can be made that Bradley was a thoughtful and far-sighted botanist. Zirkle remarks: 'More can be found on hybridisation and sex in plants in Bradley's works than in the botanical papers of any other man of his time.' Gordon Rowley, who in 1964 edited a book of Bradley's writings on succulents, described him as 'the first champion of plant breeding'.

The fact that Bradley's and Fairchild's experiments and discoveries appear to have made little immediate impact on the theory and practice of British gardening is blamed by Rowley on Linnaeus, who diverted the attention of botanists to the naming and classification of existing plants rather than to the creation of new ones. And some of those who did take notice were sceptical. In 1726 John Laurence, rector of Yelvertoft, wrote *A New System of Agriculture* in which he made fun of Bradley's claims on hybridisation: 'Doubtless Mr Bradley, having since this first conjecture had many years wherein to try his experiments and to improve vegetables, will quickly give us a history of his success and a catalogue of new fruits as well as new flowers,' he wrote with heavy irony.

The developments in the science of breeding were noticed across the Channel in France, where they were described in an odd book called *Spectacle de la Nature*, published in Paris in 1732, three years after Fairchild's death, and in an English translation in 1740. The relevant passage took the form of a dialogue between a count and a cavalier, who, in giving some hints on gardening, quoted 'an English gentleman' who had passed his way:

> He assured us he was persuaded that the powders which fall from the tops of the chives [stamens] are frequently wafted to some distance by the motion of the air; and that by their action on the pistil, or style, of another flower of the same species, but of a different colour, they impregnate some of its seed and diffuse a new tincture into the colours of the flower, which springs from that seed.

The cavalier went on to describe Bradley's experiment with the castrated tulips and the isolated auriculas before coming to Fairchild's mule:

> He surprised us extremely when he added that the same experiment made on flowers which entirely differed from each other in their nature and qualities produced seeds; and that the flowers which sprang from them were composed of those different qualities. But he assured us that these new flowers, which had no similitude to any others he had ever seen before, were unproductive of any seed the next year and not perpetuated like the rest.

The count grasped the point immediately:

> Were this fact certain, it would have some correspondence with the birth and sterility of mules, who may be considered by us as monsters because they are the offspring of animals who not only differ in species but likewise in nature . . . Though I am far from resigning my judgment to the first plausible idea presented to me by other persons, yet I think it is a criminal presumption to be so tenacious of one's knowledge as to dislike any mention of new discoveries. We are still in the infancy of arts.

Fairchild shared that positive attitude to experimentation. By going through Bradley's voluminous writings, we can build up a picture of Fairchild as a man constantly seeking to try something new and pondering the implications of the phenomena he observed. He amassed a collection of more than a hundred plants with variegated foliage, so as to confirm his theories about the circulation of sap – presumably by grafting variegated material on to plain and seeing where the striped leaves first made their appearance. Jasmine was among the most popular shrubs of the time, and Fairchild developed a number of forms with variegated leaves, although some species are not easy to graft. He produced oranges and lemons on the same tree and black and white grapes on the same vine.

The distinguishing characteristic of the 'curious' gardener was a desire to know not just how plants and trees reacted to certain treatment but why. Both Fairchild and Bradley fell into that category. In his *General Treatise*, Bradley discusses why fruit trees, roses and flowers planted just outside the wall of Fairchild's stove were much more forward than elsewhere in the garden: 'The heat of the wall, I believe, did not only make them push their blossoms forwarder than ordinary, but contributed to make them grow in their root and get nourishment as soon as they were planted.'

From a fawning letter to Fairchild that Bradley published in the same volume, it is clear that the nurseryman had agreed to perform experiments on his behalf: 'I write you this to claim your promise of trying any experiments that I shall direct you in writing. But I would not treat the most rational gardener I have ever met in the same manner I would an ordinary one' – just as a racehorse trainer would not presume to tell a champion jockey how to ride a race. Despite that reservation, Bradley goes on to ask the nurseryman to try some further unlikely grafts, such as almonds on to plums and peaches, to test the theory that any tree whatever can be grafted on another.

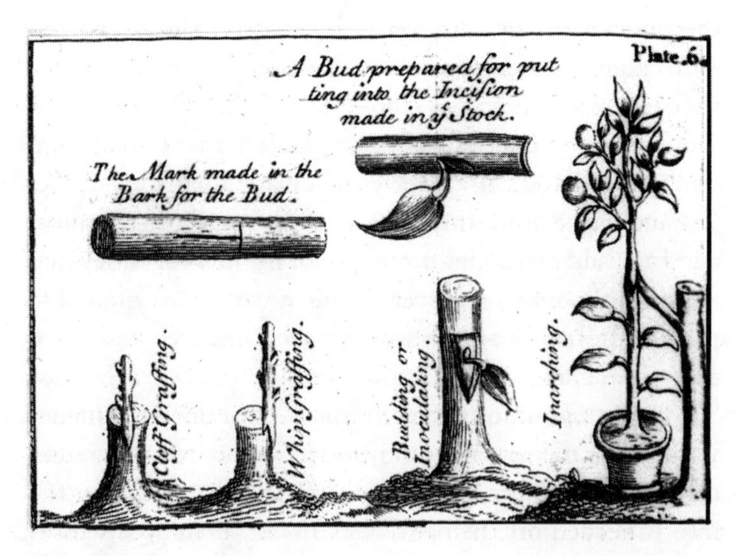

Bradley illustrates the science of grafting in his *General Treatise on Husbandry and Gardening*

A subsequent letter to Fairchild contains more fulsome flattery ('The memorandums you sent me have given me extraordinary satisfaction, as it is plain they were penn'd by a person of learning') and more instructions for attempts at crossbreeding, this time with mistletoe and flowers as well as trees. He suggests grafting a stock on to a wallflower and budding mistletoe on to a mezereon (daphne). Inevitably, most of the experiments would fail, and not for another 200 years would the science of genetics be sufficiently advanced to explain precisely why. Among those that succeeded, only a few would have been worth developing: even today breeders produce scores of poor specimens for every one that is good enough to introduce into the catalogues. But the prospect of endless novelty greatly excited Bradley and Fairchild, seeking as they were to define the limits of the possible.

By now other gardeners and botanists were devising their own experiments to determine the sex life of plants. As Blair had informed Sloane as early as 1721, Philip Miller had discovered that spinach, like the date palm, was either male or female, not bisexual, and that the females would produce fertile seed only when males were in the vicinity. He was also among the first to observe the role of bees in spreading pollen to receptive pistils, describing his discovery in letters to Blair and Bradley.

Yet there was still no full realisation of what the new discoveries meant to the growing band of hobby gardeners in Britain and throughout the world. That is often the case with what the press today calls 'breakthroughs' in science: time is needed for their real importance and impact to be weighed. It may be that the ability to clone animals and even humans will change our concept of reproduction – or it may be a diversionary sideshow. Genetically modified foods may solve world hunger or they may seriously damage the environment.

While we are still uncertain of the long-term outcome, our instinctive defence mechanism is to poke fun at the discoveries. Just as today's newspapers and magazines are full of cartoons about cloned sheep and Frankenstein food, so the emerging breed of eighteenth-century satirists, led by Alexander Pope and John Dryden, seized eagerly on the concept of sex in plants as a target for their wit. Moreover, erotic literature – or pornography, to put it more bluntly – was starting to come into fashion, initially in the form of translations from the French. John Cleland's *Fanny Hill*, although not published until 1748, is believed to have been circulating among his friends in manuscript in the 1730s. The notion that plants should experience a form of sexual coupling was so novel and intriguing that a number of ribald satires were published, poking elegant fun at the findings.

In 1732 a satirical poem called 'The Natural History of the Arbor Vitae, or the Tree of Life' was printed as a pamphlet for J. Wilkinson, 'near Charing Cross'. The title page said it was 'addressed to the ladies by a member of the Society of Gardeners'. By the standards of the time it was, in parts, fairly hard-core stuff. Its dedication, 'To the Fair Sex', clearly announced its intention:

> . . . 'Tis the description of that tree
> Which constitutes posterity,
> And consequent gives life to all
> The denizens of this round ball.
> You'll see, where'er 'tis found to grow
> 'Tis you its virtues do bestow:
> You are its cause, its influence;

> You its perfections do dispense.
> Thus, as 'tis raised to life by you,
> This work of natural right's your due;
> And gratitude demands likewise,
> That the whole sex should patronise.
> Then be the Tree of Life your care
> And use it gently; as you're fair;
> Cherish the plant with kindly dew,
> As it does often cherish you.

Like everything else in the poem, this was highly suggestive, and in case readers did not get the point it was spelled out a little later:

> The TREE OF LIFE, another name
> For P-n-s, but in sense the same,
> Is a rich plant of balmy juice
> And own'd to be of sovereign use,
> Consisting of one single stem
> That's straight, as is a Pistillum;
> Whose top sometimes the curious say
> Is like a cherry seen in May;
> Or glandiform – but's found to be
> More oft like nut of filberd-tree.
> Quite the reverse of other fruit,
> This grows, and dangles near the root,
> Producing two of nutmeg kind.
> Twin-like, in one strong purse confin'd . . .
> The fruits receive a strong supply
> And yield a viscous balmy juice
> Adapted to VULVARIA's use:
> That is, from time to time, it flows
> At the pistillum, and bestows

Itself, in chief, on th' open cup
O' th' flowering shrub, which drinks it up.

When it comes to naming names, the usual suspects make their appearance. 'Th' ingenious Bradley' is credited with the revelation that, after fertilisation, plants 'commence pregnancy'.

Thus the Vulvaria's proved by art
To be the other's counterpart,
And so it may, with property
Be call'd, of Life, the Female Tree . . .
All climes this plant produce, tho' some
Thrive more, or less, in Christendom,
And do increase to larger size
As under more propitious skies.
Nine inches we can rarely boast
In England; or eleven at most.

If the climate is unsuitable, artificial heat must be applied:

These trees don't raise with ease their heads
In winter, without good hot beds:
In warmer weather they can bear
To stand, expos'd to open air.

Care must be taken to avoid the infections and infestations that so often accompany this pleasurable pastime. Readers are warned against 'the morpim breed', a reference to body lice, or crabs. It is easy to mistake a foul vulvaria for a wholesome one – and here is added a note that is doubtless a topical reference: 'I perceive Mr Bowen is perfectly acquainted with the French . . . value of experience, since he so dearly paid for it.'

The poet warms to the theme:

> These ugly, loathsome maladies
> Which so affect such useful trees
> Have ages exercis'd mankind
> A proper remedy to find . . .
> These venomous vulvariae
> Spread wide the great variety.
> Search but this spacious town about,
> You'll find few gardens it without.
> Travel St James's Park around,
> There too in plenty they abound:
> And in whole shoals, but cast your eye
> Upon the gardens that do lie
> On t'other side o' th' Thames – I call
> Not Vaux but Rottenvulva's hall.

Vauxhall Gardens, on the Surrey side of the Thames, had been a popular resort for Londoners since the Restoration. It was noted for its music, for the meagre portions served by its caterers and for its abundance of loose, unhygienic women.

A prose version of the poem, containing roughly the same jokes, was published as a pamphlet in the same year. In 1741 it was republished along with a new and even more ribald prose satire: *Natural History of the Frutex Vulvaria or Flowering Shrub, as it is Collected from the Best Botanists both Ancient and Modern* by Philogynes Clitorides. The dedication, to 'the two fair owners of the finest vulvaria in the three kingdoms', speaks of various calamaties that have befallen the nation in the last hundred years and declares: 'All these misfortunes the naturalists and botanists ascribe (how truly

I know not) to the degeneracy of our Tree of Life; how much then, beauteous ladies, must the whole nation be obliged to your indefatigable endeavours to restore their vigour.'

This botanical satire on the female organ is cruder than its earlier companion piece: there was clearly a limit to the hilarity that could be squeezed from comparing the naughty bits of plants to the naughty bits of people. It begins: 'The Frutex Vulvaria is a flat, low shrub, which always grows in a moist, warm valley at the foot of a little hill, which is constantly water'd by a spring.' Where the Arbor Vitae was valued for its size, vulvarias were best when small – and quite useless if their diameter exceeded five inches. There was a reference to the rabbit lady of Godalming, exposed by James Douglas, and a number of jibes about men and women whose names, thinly concealed, mean nothing to readers more than 250 years on. The jokes about venereal disease and prostitutes surface again: the vulvaria – said to have been most prized in the reign of George I – thrives in 'hot-houses', especially in the vicinity of Drury Lane, St James's, Westminster, St George's Fields and Vauxhall – all then notable for their brothels.

A less vulgar verse, referring directly to Fairchild's mule, was written in 1781 by Erasmus Darwin (Charles's grandfather) in his *Botanic Garden*:

> Caryo's sweet smile, dianthus proud admires
> And gazing burns with unallow'd desires;
> With sighs and sorrows her compassion moves,
> And wins the damsel to illicit loves.
> The monster-offspring heirs the father's pride.
> Mask'd in the damask beauties of the bride.

Caryo is a reference to *Dianthus caryophyllus*, the botanical name of the carnation. The monster-offspring – shades of Frankenstein foods – is the mule.

At the age of 56, after a decade and more of rubbing shoulders with men formally trained in science, Fairchild felt confident enough to accept an invitation to address the Royal Society in his own right, rather than have his work described by a Fellow. As he made the journey to Crane Court for the second time, on 2 April 1724, struggling with an even larger bundle of samples than he had taken along four years earlier, he could have been forgiven for reflecting, with just a tinge of self-satisfaction, that this was the high-water mark of his career. What is more, this time Sir Isaac Newton was to be in the chair: at 82, though increasingly frail, he was far from senile and still liked to keep up with the latest discoveries. It is true that the topic of Fairchild's paper was not as ground-breaking as that which had occasioned his last visit to the Society, but it was one that never failed to excite botanists – the circulation of sap.

Apart from the nurseryman himself, the only other non-Fellow of the Society to attend was Philip Miller – described in the minute book as 'the king's gardener', although he had been at the Chelsea Physic Garden for two years. After the preliminaries (including the re-admission as a Fellow of Sir James Thornhill, painter of the marvellous murals at the Royal Naval College at Greenwich, who had let his membership lapse), the first speaker was John Brown, a chemist, who described a method of manufacturing the colour Prussian blue. Next came Gabriel Daniel Fahrenheit, no less, the eminent German physicist who gave his name to the temperature scale still widely used: his paper this time was on freezing water in a vacuum. Fairchild was last, topping the bill even in such distinguished company.

His original manuscript is preserved in the Royal Society's archives, and two copies are in the British Library. He had small, neat handwriting, to be expected in a master of a trade where precision is paramount. There are several crossings-out, most of them serving to soften his conclusions, making his argument more tentative and less didactic. Perhaps he showed the draft to one of his more learned friends – Sloane, maybe, or Blair – who told him that the conventional form of presenting scientific argument was as supposition rather than assertion.

Thus, the first alteration came in the title of the paper. The original 'account of some new experiments to prove the continual circulation of sap in plants and trees' was changed to the non-committal '. . . experiments relating to the motion of sap . . .'. Later he described an experiment on grafting that showed branches of a Virginia cedar (another Catesby contribution?) growing both above and below the graft. In the original text he asserted that this 'proved' the circulation of sap. In the event he deleted the word 'proved', commenting merely that the result was remarkable.

He began the talk by recalling his first visit to the Society some four years earlier. The Fellows 'were pleased to allow the experiments [that I then showed them] to be new and useful, which encouraged me to try further and bring more experiments in order to show the course of the sap'. Through acting on his findings, 'I can make barren trees fruitful and decaying trees healthful, and render the system of gardening and planting more useful to the public'. He noted that evergreen trees grafted on to deciduous root-stock held their leaves in winter, proving that the sap from the deciduous roots rises in winter, otherwise there would be nothing to nourish the leaves. He suggested that if foreign oaks – which tend to be spongy – were grafted on to firm English stock, they might produce more solid timber than trees raised from sowing foreign acorns.

To confirm his theories he had brought along four specimens from the twenty-one grafts he had listed in Bradley's *General Treatise* a year earlier. His selection of which ones to show the Fellows appears to have been based as much on their crowd appeal as their intrinsic scientific worth. The first, the cedar graft, was straightforward enough, but he followed it with a viburnum that had been planted upside down, with its branches in the soil: the branches became roots and the roots branches. He was not quite sure what to make of this:

> Whether the same vessels which fed the branches have changed their course or whether the juices go up and down in the same vessels I shall leave to better judgments, but I find the plant in as good a state of growing as it was in its natural state.

His third sample was a pear tree that he had grafted on to rootstock three years earlier but had left out of the ground since then. As he was able to show the Fellows, it had survived without being in the soil, with a branch in blossom and suckers shooting from the root, 'which proveth that the branches are as useful to support the roots as the roots the branches; and it is therefore no wonder that so many trees miscarry in planting, when there are no branches left on the head to maintain circulation in the root'. Finally, he showed an evergreen cedar of Lebanon that had held its leaves all winter despite being grafted on to a larch, one of the few deciduous conifers.

The final paragraph of the original paper has been crossed out. It read:

> I have several more examples and shall continue to go on by way of experiments, but hope these will be sufficient to prove the circulation of the sap, which will be very useful to the public, which is the hearty wishes and sincere desire of your most humble servant, Thos. Fairchild, Hoxton, Apr: 2, 1724.

There is evidence that this date has been altered, possibly from 4 July 1723. This suggests that the original paper may have been in the form of a letter to Sloane or some other Fellow of the Society, describing the points he planned to make in his talk, and that when he came to make his appearance he simply read the letter to the gathering, with slight alterations. When he had finished he was thanked by Newton, who ordered that his paper be printed; so he left it with an officer of the Society before proceeding, no doubt, to another evening of botanical talk around the dinner table of a City inn.

After a lifetime of dedicated application to his craft, Fairchild's fame was now at its zenith, and it must have been at about this time that he had his portrait painted. The picture survives in the Department of Plant Sciences at Oxford University, hanging at the top of a staircase not far from the Sherard Herbarium; but there is no clue as to why or for whom it was painted. It shows a chubby, clean-shaven man in his fifties, with a benign, open face, a half-smile playing on his lips, his left hand cupped around his ruddy cheek. He is wearing what seems to be a modest shoulder-length light-coloured wig, well combed and gently curled – not one of the long periwigs that had been fashionable in the previous century. His jacket is open and a white cravat is tied around his neck. In his right hand, resting on his knee or the arm of a chair, is an open manuscript with diagrams, almost certainly a botanical work of some kind.

The portrait is attributed to Richard van Bleeck, a Dutch painter born in 1670 who spent much of his working life in England. His most celebrated sitter was the playwright William Congreve, also depicted holding a manuscript. Among his aristocratic patrons were the Howards, England's senior Catholic family, and he painted several pictures of the eighth Duke of Norfolk. Other notable portraits are of Sir John Holt, the powerful judge, the dissenting minister John Guyse and Selina, Countess of Huntingdon. In the Clothworkers' Hall in London is a van Bleeck portrait of Sir Robert Beachcroft, Master of the company from 1700 to 1701 and Lord Mayor of London from 1710 to 1711. This suggests that the painter had links with the guilds, so the Fairchild portrait could have been connected with his membership of the Gardeners' Company – although the company itself would have been unlikely to commission it, having no hall to hang it in. It is quite possible that he commissioned it himself: Samuel Pepys had himself and his wife painted twice, for sums ranging between £1 and £5.

By 1756 the portrait was in the Ashmolean Museum in Oxford. A note in the museum catalogue, published that year, says that it was donated by Charles Moore of St John's College. In 1897 it was transferred to the offices of the Oxford Botanic Garden and then to the Department of Plant Sciences. Charles Moore, born in 1722, was at St

John's between 1740 and 1745, practised as a lawyer and became a Fellow of the Royal Society in 1768. How he acquired the portrait and whether he had any family or other links with Fairchild are unknown. St John's was the former college of William Sherard, who bequeathed some paintings to the university, but it seems unlikely that he would have possessed a portrait of Fairchild: while they certainly knew and respected each other, and Sherard had sent some of his plants to be nursed at Hoxton, there is no other evidence that they were especially intimate.

Like the dedication of *The City Gardener* to the governors of Bethlem and Bridewell, the purpose of the portrait is one of the mysteries surrounding Fairchild that may never be solved.

CHAPTER SEVEN

Death in Hoxton

The vegetable world was his province. It was his study, as well as his maintenance. He every day saw and admired its beauties and adored the Great Author of such admirable, such inconceivable perfection. He, like our one happy parent and progenitor, lived in a garden. But happier far: he kept possession, kept his innocence. Not content with enjoying such felicity, all his life long, he gave his substance to perpetuate that pleasure in others.

REVD WILLIAM STUKELEY, GIVING THE ANNUAL FAIRCHILD LECTURE AT ST LEONARD'S CHURCH, SHOREDITCH, IN MAY 1760

On 20 March 1727 Sir Isaac Newton died, aged eighty-four, and eight days later he was buried in Westminster Abbey. During a rancorous seven-month campaign for the succession to the presidency of the Royal Society, factional rivalry reached fever pitch and old friendships were called into question: Bradley, typically, was accused of deviously promising his support to more than one candidate. Eventually, in October, Sir Hans Sloane was elected. In June, Fairchild had celebrated his sixtieth birthday, still reasonably robust but naturally less sprightly than he once was. His nephew Stephen Bacon, although barely out of his teens, was being groomed as his successor and assuming more and more of the day-to-day responsibility for running the nursery. On 11 June, a few days after Fairchild's birthday, a new reign began as George I died in Hanover.

It was a warm, dry summer, culminating in the coronation of the new king on 11 October. The weather was good for gardening and good for business. The following year, too, began propitiously. On 1 March 1728 the first flowers appeared on the *Lilio narcissus*, a South American lily, probably a species of amaryllis, whose bulb had been put in his care by James Douglas three years earlier. As it happened to be the queen's birthday, Douglas decided that the flower should be renamed *Lilio narcissus reginae*, and he reported on its flowering almost immediately to the Royal Society, taking a sample with him to its 7 March meeting, as well as a root of the fashionable Guernsey lily for the sake of comparison.

Of the *Lilio narcissus*, he said:

> The root as Mr Fairchild informs me is of the bulbous truncated kind, made up of several parts encompassing one or other in the same manner as may be observed in that of the Guernsey lily, which lies here before me . . . The flowering plant has at present the leaves which began to shoot out before Christmas, at which time it had four withered leaves which Mr Fairchild cut off upon the appearance of the new ones, and he informed that it has never been bald of leaves since the time it first had any . . .
>
> About the history of this beautiful plant, it may be worthwhile to observe that the roots of it were brought by a gentleman from Buenos Aires about three years ago, having been some time before carried thither from Mexico, the original *locus natalis* thereof . . . When the roots arrived in London they were very much wasted and decayed, but falling by good luck into the hands of the most ingenious and skilful Mr Fairchild of Hoxton, he with surpassing care and pains recovered them. A surgeon could not

> have treated a mortified animal member with more judgment and dexterity than he showed in his management of this corrupted vegetable. He cautiously separated the mortified parts from the sound and, by the application of artificial heat, cherished and recovered the small remains of life, and thus by degrees brought them to that healthful and thriving state in which we now behold them.

Sadly, though, the health of the lilies was not matched by Fairchild's own condition, nor that of some of his acquaintances. A generally wet summer put him in poor spirits, and in August he learned of the death of the botanist William Sherard, for whom he had nursed a number of tender plants. Sherard left a substantial estate of £20,000, including an endowment for the chair of botany at Oxford. Two months later Fairchild was mourning Dr Thomas Bennet, the parish priest at St Giles, Cripplegate – his reserve choice as the venue for his memorial sermon. They appear to have been good friends, for in *The City Gardener* Fairchild had described the advice he gave to Bennet on pruning his figs.

The damp summer was followed by a severe winter, with frost and snow from mid-December until the end of January. The weather improved as spring gave way to summer, but Fairchild's health did not. It was clear to his friends and associates that he did not have long to live. The exact nature of his ailment is unclear, but by 21 February 1729 he felt it advisable to draw up his will – a sombre task that in those days would be undertaken only when it seemed certain that life was starting to ebb away. Mark Catesby, then working at the nursery, was one of three witnesses to it; the others were Elizabeth Seamark and John Brookes, perhaps neighbours or domestic servants.

Eighteenth-century wills were elaborate documents, composed to a formula. 'In the name of God, Amen,' this one began:

> I, Thomas Fairchild of the parish of St Leonard Shoreditch in the County of Middlesex, gardener and citizen and clothworker of London, being in good health of body and of sound and perfect mind and memory (praise be therefore given to almighty God), but considering the uncertainty of this present life, do make this my last will and testament.

Having recommended his soul to God, asked for the forgiveness of his sins and the inheritance of eternal life, he made the first of the will's several unusual provisions: 'My body I commit to the earth to be decently buried, at the discretion of my executors hereafter named, in some corner of the furthest churchyard belonging to the said parish of St Leonard Shoreditch, where the poor people are usually buried.' This referred to a piece of land by the Hackney Road, separate from the churchyard, bought by the church in 1625 as a subsidiary burial ground. It was an odd request that can be interpreted only as evidence that, despite his worldly success in life, he wanted to be identified in death with the yeoman stock to which he was born.

There followed a list of bequests to relatives which show that he had no direct heirs. Most of his relatives and beneficiaries were either from his mother's family or descended from her third marriage to John Bacon. Fairchild's cousin Richard Butt, who would receive £30, was the son of his mother's brother. (A Richard Butt owned a nursery at Kew between 1731 and 1751, but there is no firm evidence to connect the two.) Stephen Bacon the elder, whom Fairchild described as his 'brother-in-law', was in fact his half-brother, the son of his mother and John Bacon. He was to receive £20, plus another £40 for each of his two daughters, to be held in trust for them until they reached the age of 21.

Stephen Bacon's sons, Stephen and John, received more substantial bequests. Stephen was the elder of the two, but both were evidently under 21 at the time the will was made. John was left £200, half to be paid when he reached 21 and the other half at 24. John was also to receive any money due to Fairchild for his contribution to the Society of Gardeners' *Catalogus Plantarum*, which would be published in 1730. The nursery and the rest of the estate were to go to Stephen. If either brother died before Fairchild, the other would take his share. (In the event John, the younger brother, died a few weeks after Fairchild, aged 17.)

Fairchild also wrote in a clause exempting his half-brother Stephen Bacon from any outstanding debts to him, suggesting that in his lifetime Fairchild had been generous in supporting his family. Thomas, Ann and Lydia Woodley, two nieces and a nephew, would inherit £40 or £50 when they came of age. They were the children of Fairchild's half-sister Ann, another of John Bacon's children, who married John Woodley in 1700.

The next bequest is the most puzzling. Fairchild left £30 to 'my daughter-in-law Mary Price, the wife of James Price'. Daughter-in-law could at that time mean a step-daughter as well as the wife of a son, but either construction would require Fairchild to have been married at some stage in his life. There is no mention of a wife in any contemporary reference to him, and if he remained a bachelor it would not have been all that unusual: Earle says that in the late seventeenth century about a quarter of the English population never married.

One possibility is that Mary Price was a grand-daughter of John Bacon and his second wife Mary, whom he married in 1680 after the death of Fairchild's mother. She would not therefore have been a blood relation: technically, she would have been a step-niece, but as no such word then existed, it is plausible to assume he may have called her a daughter-in-law. Other scenarios would have Fairchild married briefly to a woman who already had a daughter, but who may have died soon after their marriage, or that Mary was a young relative whose parents had both died and for whose welfare he had assumed responsibility. If the reference were to be taken in its present-day meaning, it would mean that Fairchild had a son, who survived to marry but then died, and that his widow Mary remarried. The lack of any other evidence of a son makes that explanation improbable, though not impossible.

There was a bequest of £20 to 'my cousin Susanna Towell, the wife of Thomas Towell'. Richard Butt, Fairchild's uncle (father of the Richard Butt mentioned earlier), had a daughter named Susanna, and this might have been her. Then comes £5 to 'my late servant John Sampson' and £15 to 'my servant Anne Taylor'. The implication of 'late' is that Sampson did not work for Fairchild at the time the will was made, but he (or his son) presumably returned to the nursery and briefly inherited it, for a John Sampson is recorded as having paid the poor rate for the property in 1737, after the death of the young Stephen Bacon. Ann Sampson, presumably John's widow, paid the rate in 1740, before the nursery was wound up.

Then came a £10 bequest to the charity children of the parish of St Leonard's. Daniel Defoe, in his *Tour through the*

Whole Island of Great Britain in 1726, said that there were 20,000 people supported by charity in London, not including the impoverished children whose schooling and apprenticeships were paid for by charitable donations. 'There is no city in the world can show the like number of charities from private hands,' he declared proudly.

The charity school movement began at the end of the seventeenth century, under the auspices of the Society for the Promotion of Christian Knowledge (SPCK). It attracted financial support from devout churchgoers because the schools were seen as a Protestant bulwark against what was still regarded as the very real menace of Roman Catholicism. In his book *The Charity School Movement*, M. G. Jones wrote: 'By the end of the first quarter of the eighteenth century, support of the charity schools was the favourite form of practical piety in London, and it is clear that the schools were objects of pride to its citizens.'

By 1704 there were 54 schools in 32 parishes with more than 2,000 pupils; 25 years later that had risen to 132 schools and 5,225 pupils. The St Leonard's School opened in 1705 in a room in Pitfield Street, Hoxton, and moved a few years later to larger premises in Kingsland Road, where it catered for 50 boys. In 1709 a similar school for girls was opened. Tributes paid to Fairchild some years after his death stress his support for the local charity school, although he does not appear in its list of benefactors. Either he gave anonymously, as was quite common, or his support went no further than that single bequest.

The will then turned to some smaller legacies. There is £5 'for mourning' to 'my friend Richard Spier of Hoxton, gardener'. Richard Spier or Spires had run a nursery close to Fairchild's since 1724, and it stayed in business for most of the century. A guinea was bequeathed to Mark Catesby to buy a ring – this would also have been for mourning, as it was the custom to leave money for mourning rings to friends and relatives. Other guineas would go to a servant, Stephen Best; 'my brother-in-law John Bacon' (again, a half-brother from his mother's third marriage) and his wife and three children; and a further guinea, plus 'all my wearing apparel both linen and wools', to John Fairchild, a cousin. No doubt this was the John Fairchild who was apprenticed to the clothworkers in the 1690s: assuming he eventually took up the trade, he would have been well qualified to adapt the clothes for further use.

The next legacy was the one that would ensure the survival of Thomas Fairchild's memory:

> I give and bequeath to the trustees of the Charity Children of Hoxton and their successors and the churchwardens of the said parish of St Leonard Shoreditch and their successors the sum of twenty-five pounds to be by them placed out at interest for the payment of twenty shillings annually forever for the preaching of a sermon in the said church of St Leonard Shoreditch by the vicar of the said parish or such other person as the said trustee and churchwardens and their successors think proper in the afternoon of the Tuesday in every Whitsun week each year on the subject following, (viz.) The Wonderful World of God in the Creation or on the Certainty of the Resurrection of the Dead proved by the Certain Change of the Animal and Vegetable Parts of the Creation, and in case default shall be made in the preaching of the said sermon at the time aforesaid, then my will is that the said sum of twenty-five pounds shall be forfeited to the churchwardens of the

> parish of St Giles Cripplegate London to be by them and their successors placed out at interest for the preaching of the said annual sermon in the parish church of St Giles Cripplegate London on the subjects and in the manner aforesaid, by such person as the said churchwardens and their successors shall think proper.

This was not an unusual bequest. In his *London in the Eighteenth Century*, Sir Walter Besant wrote that almost every church in the metropolis had its 'gift sermons' that in some ways corresponded with bequests to charities and, later, to the foundation of almshouses. This was when 'all the churches were built for preaching houses': the 108 churches that he lists from the period had 189 endowed sermons between them. The essential difference between Fairchild's sermon and most of the others is that his has survived for 270 years.

The younger Stephen Bacon and Richard Spier were appointed executors of the will, which ended with a resonance and flourish characteristic of the period.

> In witness thereof to this my last will and testament contained in three sheets of paper, to the first of which I have subscribed my name and to the third and last my name and seal, the one and twentieth day of February one thousand seven hundred and twenty eight and in the second year of the reign of our Sovereign Lord George the Second by the grace of God king of Great Britain, Thomas Fairchild. Signed, sealed, published and declared by the said testator Thomas Fairchild to be his last will and testament in the presence of us who subscribed our

> name and witnessed in the presence of the said testator: Elizabeth Seamark, Mark Catesby, John Brookes.

As Fairchild lay dying, his rest must have been disturbed by an extraordinary event that occurred in the neighbouring nursery that had once belonged to his friend William Darby but was now in the hands of John Cowell. Cowell described the amazing scene in his book *The Curious and Profitable Gardener*, published in 1730.

Aloes, or agaves, with their fleshy leaves and lush habits of growth, were among the most sought-after of the plants being brought back from the tropics by plant-hunters. Some have erratic flowering habits, so that when they did come into bloom, then as now, they were much admired and marvelled at. They were among the specialities of Cowell's nursery, and the whole of the first chapter of his book was devoted to them.

'There have been so many different accounts published of the curiosities which blossomed in my garden,' he wrote, 'that a stranger cannot be certain which of them is the best, or whether any of them are true. I am therefore desired to publish my own account of the aloe and torch-thistle, as they really appeared, and give the history of them.'

He begins by discussing the 'large common American aloe', grown in England since Elizabethan times and mentioned in Parkinson's 1629 book, then popularly known as Parkinson's 'theatre of plants'. He attributes the plant's introduction to Sir Walter Raleigh and Sir Henry Carew 'who likewise were the first gentlemen who made tobacco and the orange tree familiar to us'. It was first known as the Persian aloe, because merchants trading in Persia had reported seeing it grow there. 'It must, however, according to the botanists, be allowed a native of America, from whence it was first brought to Europe; and was so agreeable to the Spanish climate that it grows there in the natural ground, without any shelter.'

This aloe was called the century plant, because it was reputed to flower only once in 100 years, although today most of them manage to sport a bloom at intervals of between 10 and 30 years. By repute, the first to flower in Britain was in Mr Versprit's garden at Lambeth, said to be one of the originals brought in by Raleigh. If that was so, it would have been 100 years old when it flowered, on stems some 20 feet tall, during the reign of William III. Cowell writes:

> The account I have of this plant is that it shot forth a long and large stem about August, which branched like a tree, and every branch contained a large cluster of blossoms and buds. The season of the year being towards a time when exotic plants required housing, made Mr Versprit build a glass case on purpose for its shelter; but either through the fault of the workmen, or the extraordinary high winds which followed, the glass case was blown down and the flowering stem of the plant broken from the root.

Two aloes, apparently also from Raleigh's batch, bloomed at Hampton Court a few years later. Cowell reports:

> When these flowered they were in extraordinary strength, even so great that one of them brought five and the other seven stems of flowers: every stem branched like a regular tree, and each branch was loaded with blossoms which, as Dr Bradley says,

> dropped honey in abundance from every flower. He further relates that they lasted the greater part of the winter in blossom, though they were not housed, and even the spring following attempted to make new branches, though the plants in their leaves began to rot towards their roots, and in their decaying state were thrown into the Thames.

He reports flowerings in Prince Eugene's garden near Vienna and the Duke of Buckingham's in London. One of the earliest flowerings in Europe was described by a French writer in the mid-seventeenth century – an event that highlights the perils of relying on eyewitness accounts of botanical, or indeed any phenomena, especially when they are written in a foreign language: 'That French author gave occasion to the common opinion that the aloe at the time of opening its blossom made a report as loud as a cannon; but the mistake was by the translator.' When the Frenchman used the words 'La plante faisait un si grand bruit' when it flowered, he did not mean 'bruit' in the literal sense of a noise, but in the metaphorical sense of a public commotion or sensation. 'This construction is allowed by the critics to be good, since the blowing [flowering] of my aloe has drawn so many thousand persons to see it from all parts of the kingdom, and even from foreign countries, as well as the compliments it has been paid by the French and Dutch news writers.'

British news writers, too, got wind of the story. The *Daily Journal* of 26 August 1729 carried a report of the phenomenal crowds, drawn by the simultaneous blooming of the aloe and the night-flowering torch-thistle. 'Physicians and learned gentlemen', it wrote, had gone in droves to Cowell's garden and stayed all night 'to see the serus or torch-thistle, which blooms when the sun goes down and closes at sunrise'. By 10 p.m. the purple, yellow and white flower was fully open. At 7.30 a.m. it started to close, and the flower had disappeared by 10 a.m. Four days later the paper reported another night-time visitation from scientists bearing microscopes, with artists on hand to record the event.

The alarming outcome of all this publicity was reported in the *Daily Journal* of 8 September. There was a riot at the nursery, in which Cowell was badly injured by visitors trying to seize his plant. In his book, Cowell describes the commotion in quite horrifying detail, first giving a full account of the flower's development during that spring and summer:

> At length my own great aloe began to open its crown for flowering; that is, its central leaf, or middle spire, as some gardeners call it, began to split and crack, and the centre of the whole plant swelled to a great degree towards the spring of the year 1729. And about the beginning of June following, the pointed leaf in the middle of the plant quite opened itself and smaller leaves appeared, to the number of six or seven, which is a certain mark of flowering; for this was observed on 8th June and the blossom bud appeared above the plant on the 10th of the same month.
>
> The bud appeared at its first shooting like the bud of an asparagus, about 13 or 14 inches in circumference, and rose gradually higher and higher, growing about seven inches taller every day at first, and afterwards about five or six inches in a day; decreasing still in its growth as it arrived nearer to its intended height; spreading into branches as the stem mounted, till the whole of the flowering stem was 20 feet high, and 17 inches in circumference at the bottom.

It carried on like that for seventeen weeks, then stopped growing for three weeks while the flower buds formed. When they did, Cowell estimated that there were around 110 blossoms on each of 30 branches, making 3,300 in all. The flowers had yellow petals about three inches long, with a two-inch pistil in the centre, surrounded by six stamens and apices which, when opened, were covered with yellow dust. 'Each flower contained near a teaspoon of sweet juice, like honey, which rose in so great quantity as to overflow the petals or divided leaf of the flower. I saved some of the liquor in a phial, which in a few days became as foetid to the smell as the liquor fresh gathered was sweet to the taste.'

He goes on to relate the history of his aloe, which he inherited on taking over the nursery in 1718. (The local rating records suggest that Cowell moved to Hoxton from a nursery on Kingsland Road in Shoreditch, close to St Leonard's Church.) William Darby had reputedly bought the plant from Mr Versprit of Lambeth soon after he established his nursery in 1677, so it would have been related to the one involved in the accident with the Lambeth glasshouse. Darby thought it was 20 years old when he acquired it, which would have made it 72 when it flowered for Cowell in 1729.

In Cowell's care the aloe had a perilous history. Some five years before the flowering, as he reveals, it received 'an accidental bruise in the shank', and as a result 'the body of the plant, between the root and the leaves, began to rot on one side, and continued to do so till there was little room left of saving it'. According to his description, he cut off the main root as well as the damaged part and left the wound to heal in the sun. Then he replanted the aloe in fresh soil 'and in a month or two my plant had shot roots that almost filled the tub; and from that time it began to grow vigorously, till at length it came to produce the blossom I have described'. He speculated that if the plant had not sustained this injury it would have grown stems half as long again, or some 30 feet high, equalling one of the aloes at Hampton Court and probably producing even more flowers.

When he observed that it was coming into flower, Cowell built an open-topped glass frame for the aloe and placed it in the south-facing part of his nursery. As the flower stem grew above the glass, he raised the height of the frame and covered it over when he observed that growth had ceased. 'I believe I may venture to say that, by this method of proceeding, the flower stem was some feet higher than it would have been had I left it exposed to the open air, as those were which I have mentioned in the Duke of Buckingham's gardens.'

He describes the plant and various legends surrounding it. Because its leaves come to a sharp point and have fibres attached, it has been called 'Adam's needle', in the belief that the first man may have used it in the Garden of Eden to stitch his fig leaves together. Cowell is of the belief that, if Adam had used a plant for this purpose, it would more likely be the yucca, which comes from Africa, 'for the learned agree that America was not known until Columbus's time, and therefore our Aloe could not be known in the first times'.

He tells us, though, that its fibres had been used as textiles more recently, quoting Bradley as his authority. The Spanish viceroy in the West Indies had some lace manufactured from them and sent samples to the king of Spain, along with some of the thread. The king apparently passed some of the lace and thread on to the French king,

Louis XIV, who in turn handed them to a French botanist, and he gave some to Bradley. 'Since which time I have got several purses made of the same aloe thread, artful enough in their work though made by the Indians, and I design to have some curious pieces dressed from the leaves of my own aloe.' Coarser fibres from the plant are used by the Indians to make fishing lines and to weave hammocks and nets 'and will bear the weather much better than hemp'.

Cowell warned readers of the caustic quality of the plant's juices: attempts to use it on the skin could result in serious blistering, but it was useful for scouring pewter and might even serve to dissolve metals. After giving directions as to its cultivation, he concluded, like any good story-teller, with the dramatic, violent climax to his tale, perhaps the first recorded incidence of plant rage. His account is so graphic that I quote much of it verbatim:

> I cannot part with this subject without taking notice of a detestable piece of malice and abuse that was offered me when this aloe was flowering in my garden, and gave me the fairest prospect of possessing an early fortune for my life, from the vast concourse of people that daily resorted to my house to see it.
>
> When the aloe was in so great perfection as to invite more company than my house and garden could well contain, and the last flower of my torch-thistle was opening, three men, habited like gentlemen, were inadvertently let up to see it: who no sooner were come to the plant but one of them began to break off the buds; and being desired to desist, took hold of the main stem and endeavoured to break it by violence; but it was luckily much too strong to give way to their base intent.
>
> This their attempt was soon discovered by all the

> gentlemen and ladies in my garden and I was called to the assistance of my servant, and to save my plant from the fury of their rage: when immediately one, who was on the top of the staircase in my house, being entreated by me to come down, fell a swearing, and drew his sword upon my man, telling him he would run him through the body if he offered to assist me; and in the meantime kicked me on the head while I offered to go up, while another at the bottom of the stairs, one of his companions, pulled me by the legs; and a third of them wounded me with his sword in two places of my neck.

As a result of this serious assault, Cowell 'was under the surgeon's hands for many weeks, devoid of attending the curious persons that did me the honour of coming to my garden'. Worse, on hearing news of the affray, those who were on their way to see the garden turned back, 'to the great loss, not only of the money I might have gained but, I fear, that noble company might be disobliged'.

Cowell said he had prosecuted those men responsible for the vandalism and violence, and he vented his anger by drawing a lengthy moral:

> I beg leave to ask, is such treatment allowable in reason, or does it not appear to be malicious, when men clandestinely endeavour to destroy the goods of a man who has got them honestly, and has showed himself with good nature to all mankind?
>
> Do not such actions carry the face of folly with them (to make the best excuse) to insult a man in his own house, whom they never had seen before? One might have expected such treatment from a madman.

> Is it not the character of men of base principles to act cowardly? And when they have a superior power, or undeserved weapons in their hands, to attack a man unarmed?
>
> Is it not villainous to destroy, or attempt to destroy a man's estate? Common robbers endeavour to get money to support themselves by robbing, but such would ruin other people out of pure ill nature. However, they have paid pretty well for that; and I hope they will be better men or good boys for the future, and not run such lengths as may bring any of them to a villainous end.
>
> I leave these sentiments to be considered by the true gentlemen, and appeal to them whether such transactions can be allowed warrantable by those who are generous enough to be just.

Fairchild would have heard the commotion as he lay on his sick bed only hundreds of yards away, knowing that he did not have long to live. No doubt he was horrified at the injury to his neighbour, but intrigued when told what the fuss was about. Within his lifetime, he might have reflected, flower gardening had developed from being the exclusive preserve of the exceedingly rich to a popular enthusiasm capable of drawing large, curious and excitable crowds to see the latest sensations. His little experiment with the carnation and sweet william, though its full implications were as yet unappreciated by his contemporaries, would help to set in motion a whole new industry that would eventually provide these enthusiasts with an endless supply of novelty.

A few weeks later, on Friday 10 October, Thomas Fairchild died, aged 62. The will was proved the following Monday, and two days later he was buried, as he had wished, in the ground reserved for the Shoreditch poor. The *Daily Journal* carried this combined report and obituary:

> Last Wednesday night, about 9 o'clock, was interred at Shoreditch the corpse of the noted Mr Thomas Fairchild, gardener at Hoxton, who died the Friday before of a palsy [a general term for paralysis]. The funeral was performed in a very handsome manner: the body lay in state at his house and the charity children, both boys and girls, to whom he bequeathed ten pounds, sung before the corpse. The pall was supported by six of the Society of Gardeners and there was a great number of mourning coaches and a multitude of wax branch lights [triple candlesticks used for devotional purposes].
>
> He was a very industrious and ingenious man in his profession, and eminent for several improvements in it, and died rich. By his will he ordered that his body should be interred in the Little Churchyard of Shoreditch, where the poor are usually buried, and in the most obscure part of it, which was accordingly performed.
>
> He bequeathed a sum of money to defray the expenses of an annual sermon every Whit Tuesday for ever, by the Lecturer of Shoreditch for the time being, on the following excellent subject, viz. The Wisdom of God in the Vegetable Kingdom, a subject of the highest importance to display the glory of God, and admirably adapted to his profession.

Within a few years, his nephews John and Stephen Bacon would be buried in the same grave. The tombstone may still be visited, in a small, rather desolate piece of open ground by Hackney Road some 200 yards from the church. According to *An Account of the Parish of St Leonard, Shoreditch*, published in 1873, the land, of which this space forms a small part, had been acquired as a secondary burial ground in 1625 but had been 'discontinued as such for the last few years on account of there being no more room'. The author records that Fairchild is buried here, adding:

> Here his remains lay for a number of years neglected and forgotten, the stone placed over his grave was concealed from view by the accumulation of dirt and rubbish until January 1845, when John Bentley and Jeremiah Long, churchwardens, caused the stone to be cleansed with care; it was found to be broken, the inscription scarcely legible . . . The churchwardens, wishing to perpetuate the memory of this good man, caused a new stone to be placed over his grave in May 1845.

The new stone records the date of its emplacement as being a year later, 1846, and it makes no mention of the Bacon brothers being buried there. In August 1891 the stone was restored afresh by the then churchwardens, Thomas Martinhill and Alfred Molloy.

Later, some almshouses were built on part of the secondary burial ground, and it is almost certain that the stone was then moved from its original site. The precise location of the grave of Fairchild and his two nephews is now lost, and in 2001, when St Leonard's is reopened after extensive restoration, the tombstone is likely to be placed inside the church. In the longer term there is a plan for the church garden – part of which was once named after Fairchild – to be redesigned in a Georgian manner, with the stone possibly incorporated into it. This would be some distance from the actual place of his burial but would provide a more fitting memorial to him than the present dismal setting.

CHAPTER EIGHT

The Vegetable Sermon

Into your garden you can walk
And with each plant and flower talk;
View all their glories, from each one
Raise some rare meditation.

JOHN REA, *FLORA*, 1665

Fairchild probably chose the Tuesday of Whitsun week for his memorial lecture because the Gardeners' Company, by its seventeenth-century charters, holds its annual 'court' for the election of officers on the Wednesday of that week. He may have calculated that scheduling the lecture on the previous evening would encourage a good attendance from the company's liverymen. The date has not always been strictly adhered to – sometimes it has been held on a Wednesday – and in 1999 it was brought forward a week to avoid a clash with the Chelsea Flower Show.

St Giles, Cripplegate, begun in 1390, is today wedged incongruously in the uncompromisingly 1970s environment of London's Barbican, with its tall apartment buildings and modern arts and conference centres. The main pedestrian access to the church is by the raised walkways that are a feature of the Barbican, with the result that most people's first view of it is at roof level, halfway up its fine, brick-topped tower. It has a notable history: Oliver Cromwell married Elizabeth Bourchier here in 1620, and in 1604 William Shakespeare is recorded as attending the baptism of his nephew. His contemporary, Ben Jonson, also worshipped here, and the poet John Milton was buried beneath the chancel step in 1674.

Although one of the City's few medieval churches to survive the Great Fire of 1666, St Giles suffered serious damage in a number of other fires, the first in 1545 and the last in the 1940 blitz. It has been well restored since and, no longer hemmed in by old shops and dwellings, its large windows let in a welcome amount of daylight, quite rare among London's churches. The fine east window had been concealed for centuries until the 1940 bombing, and it now contains modern stained glass depicting the Crucifixion, as well as an image of St Giles, patron saint of cripples, beggars and blacksmiths. A window in the north side is dedicated to the actor and philanthropist Edward Alleyn, who worshipped here. The church has historical connections with a number of City livery companies, and a ceremonial sword fixed to a pillar at the entrance to the choir bears the coats of arms of the last five aldermen of Cripplegate to serve as Lord Mayor of London.

At 5.30 p.m. on Tuesday 18 May 1999, 112 liverymen of the Worshipful Company of Gardeners and their guests assembled in the church for the annual guild service following the company's election court, held at Guildhall earlier in the afternoon. According to a long-standing tradition, the church was filled with large and elaborate flower arrangements – white Madonna lilies intertwined with purple clematis and other seasonal blooms – supplied by members of the Gardeners' Company and their wives. As the congregation took their seats, Percy Whitlock's 'Folk Tune' was played in the organ loft above the west end of the nave, where the St Giles' Singers sat.

The service began with a procession up the aisle by members of the Court of Assistants, preceded by the Spade Bearer carrying aloft the company's ceremonial silver spade with its carved dark wooden handle. They were followed by the Master, Norman Chalmers, and his two Wardens, wearing green fur-trimmed gowns and carrying posies of herbs. The Master's enamelled badge was engraved with the company's coat of arms, while around his neck hung his two-stranded chain of office bearing the coats of arms of former Masters.

The congregation was welcomed by Canon Peter Delaney, vicar of All Hallows by the Tower and Upper Warden of the Gardeners' Company, and as such next in line to be Master. After the first hymn ('Praise to the Holiest in the Height . . .') he recited the traditional bidding prayer, asking worshippers to pray 'for all the corporations, guilds and companies of the City and especially the Worshipful Company of Gardeners, and for the Master, Wardens and Livery of the company: for the agriculture and horticulture of this our land: and for all measures to advance health and wellbeing of its people'.

The prayer continued:

> To these your prayers ye shall add unfeigned praises for mercies already received, particularly for the advantages offered in this Worshipful Company by the munificence of founders and benefactors, such as was Thomas Fairchild, citizen and gardener, whom we commemorate this day.

The next hymn was 'To Be a Pilgrim' by John Bunyan, who died in Fairchild's twenty-first year. Then the Master read the lesson, the familiar extract from St Paul's Letter to the Philippians that ends: 'Whatsoever things are true, whatsoever things are honest, whatsoever things are just, whatsoever things are pure, whatsoever things are lovely, whatsoever things are of good report; if there be any virtue and if there be any praise, think of these things.' The anthem 'Jesus Christ the Apple Tree', sung by the St Giles' Singers, led into the Fairchild Lecture, given this year by the Right Revd Michael Mann, Assistant Bishop of Gloucester and a former Dean of Windsor.

After climbing to the high pulpit, Bishop Mann, like every Fairchild lecturer before him, put his own spin on the theme laid down by the founder. We know from Fairchild's book, with its stress on London's smoky air, that the quality of the environment was an issue in the early eighteenth century, just as it is nearly 300 years later; but it was almost certainly not what he had in mind when he asked the lecturers to address themselves to the wonderful works of God in the Creation or the certainty of the resurrection of the dead, as proved by changes in animals and vegetables.

> Of course [said Bishop Mann], lip service is paid to the beauty of the countryside, but that sometimes seems to apply only in so far as it provides a pleasant retreat for the town dweller. God made the world, but we need to remember that it has been the sweat and hard work of countless generations of Britons that have created the beauty of the countryside, with its gardens, fields and lanes . . . Anyone whose interest lies with the soil knows what a demanding taskmaster that can be. But we have inherited from our forebears a most precious and beautiful land, and that is our trust . . . We are God's hands in this world . . . The pursuit of excellence and the cherishing of all that is lovely depend upon us. Man can either enhance or destroy God's Creation, and which he will do will depend on what sort of people we become, and that means you and me.

Fairchild would have been comforted by the message. While the bishop, like him, acknowledged God's paramount role in the Creation, he recognised, too, that it was in man's power to alter and adapt what had been created. The bishop's message was that our motivation is the important factor – and Fairchild, for all his struggles with his conscience, knew that his own motives were benign. His twin aims throughout his life had been to enhance knowledge and, through his charitable work, to alleviate suffering. To him the survival of his lecture, despite many threats to its continuance, would be a signal of God's approval – even if, given his populist instincts, he would have preferred the message to have been distributed rather wider than among just the wealthy City gents who constituted the bulk of the congregation at the guild service.

The final hymn, dedicated to horticulture, was written for the Gardeners' Company by Michael Saward, Canon of St Paul's and a liveryman of the company. It began:

God of the garden, Eden's land,
where plants and trees brought rich delight;
where man and woman pleasure found,
fulfilled in all things, day and night.
Creator, Father, thanks we give
to you, the author of our joys,
and yet, despite these gifts, we grieve
for all that humankind destroys.

The national anthem was sung, and the service closed with the Court of Assistants processing down the aisle in the reverse order from the one in which they had arrived, in other words with the Master at the head. The gardeners went off for their annual supper, held that year at the Farmers' and

Fletchers' Hall in Cloth Fair, near St Bartholomew's Hospital. At the supper a touching little ceremony was played out. Bishop Mann was handed a leather purse containing 20 shillings from the reign of George II, obtained by a liveryman who worked at the Bank of England. This represented the sum paid to the early Fairchild lecturers in the 1730s. Afterwards the bishop gave the purse and its contents back, to be used again in 2000.

The first Fairchild Lecture – they were later to become popularly known as the Vegetable Sermons – was given in St Leonard's, Shoreditch, as Fairchild had wished, on the Tuesday of Whitsun week, 19 May 1730. The lecturer was the Revd Dr John Denne, vicar of St Leonard's since 1723, also Archdeacon of Rochester and later rector of St Mary's, Lambeth (now the Museum of Garden History). The length of the lecture – it took up 33 pages when later published in a collection of sermons – suggests that considerable importance was attached to the occasion, and it is likely that many eminent gardeners and scientists were in the congregation.

The tradition of decorating the church with appropriately spectacular bouquets of early summer flowers may have begun with this first lecture. It is pleasing to imagine that they were made up of many of the varieties that Fairchild had noted seven years earlier as being in bloom at the Hoxton nursery in May: pinks, peonies, lilies of the valley and day lilies, ranunculus, foxgloves, lilac, geranium, honeysuckle – even maybe a mule or two.

Perhaps as a tribute to one of the specialities of the nursery, Dr Denne took as his starting text a well-known quotation from the Gospel According to St Matthew: 'Consider the lilies of the field, how they grow; they toil not, neither do they spin; and yet I say unto you that even

Solomon in all his glory was not arrayed like one of these.' In elaborating on the theme, though, he ignored the recent botanical theories to which Fairchild's experiments had contributed, and clung to the view that a seed contained the embryo of a whole plant folded into a shell:

> There is an infinite variety of seeds, by which the different races of vegetables are preserved and propagated; so that they never fail to keep, without confusion, their natural order or species; God thus giving to every seed his own body . . . All the seeds and plants that ever were or shall be in the world were formed together in embryo by the word of the Almighty, on that solemn day of Creation when God said: 'Let the earth bring forth grass.'

In 1731 the lecture was delivered by the Revd Henry Wheatley and in 1732 by the Revd John Bridger, both priests with local connections. In 1733 Dr Denne, who as vicar was responsible for administering the legacy on behalf of the parish, made two decisions: first, that from now on he would himself deliver the lecture annually for as long as he was able – another 26 years, as it turned out – and secondly that the sum bequeathed by Fairchild was insufficient to reward him and his successors for their efforts.

His lecture that year elaborated on the theme of his first one three years earlier. He said that experience showed that no new kinds of plants or animals had been or could be created beyond those that God had introduced:

> The sun, with all the elements, in conjunction with the skill of man, or the powers of mechanism as exercised in the course of nature, have never yet been

> able to produce any new species, nor has any old one been lost, although individuals may have been improved and varied by an artificial bettering of soils . . . Whence it becomes highly probable that God created together all the seeds of vegetables that ever were or shall be in the world.

He paid lyrical tribute to the way in which God had arranged his horticultural creations for man's enjoyment:

> What pleasure is there in all the busy scenes of life, unless we can now and then be relieved from the hurry and fatigue of them by rural retirements and entertainments, where the vegetable world receives us with all the sweetness and freshness of air uncorrupted, which alone is often able to revive us when past recovery by all the powers of physic.

Dr Denne published the lecture as a pamphlet, and in a preface to it he argued the case for the sum left by Fairchild to be increased by public subscription:

> This legacy, you see, provides but a slender recompense for a preacher; and even that is likely to be lessened or lost, since hitherto the trustees have not been able to place out the principal money upon good security, so as to answer the yearly interest of 20 shillings.

Stephen Bacon, Fairchild's successor and executor, supported the initiative, and the subscription list was opened on 15 May 1733. The aim was to increase the fund from its original £25 to £100, to be invested in South Sea annuities – the South Sea Company had been reconstituted after the notorious speculation of 1720. James Douglas, giving a guinea, and Sir Hans Sloane, with two guineas, were among twenty-seven people who responded to the appeal, including Catherine Walpole, the wife of the first British Prime Minister. The largest donation – from Cornelius Wittenoom, a Shoreditch vinegar-maker – was only three guineas, and initially only £45.3s. was raised.

In 1746, after his fifteenth sermon, Dr Denne came to the conclusion that the parish was not properly equipped to handle the fund. He arranged for the Royal Society henceforth to act as trustees for the lecture, 'as being the most proper persons in whom to repose and perpetuate a trust so suitable to the very end of their incorporation, that of promoting the knowledge of natural things to the glory of God and the good of mankind'. To enable the fund to meet its £100 target, Dr Denne contributed £25.17s., saying that it was the sum he had earned from 15 years of lectures. The change of responsibility did not affect his monopoly on their delivery: he continued as the lecturer for another 12 years.

After the Royal Society took over, the event became for a while an important social occasion in the Fellows' calendar, the chance of a day out in the northern suburbs. The physician and antiquary William Stukeley, who had assisted James Douglas with his dissection of Sloane's elephant many years earlier and was now Rector of St George the Martyr in Bloomsbury, wrote in his diary for 4 June 1750:

> I went with Mr Folkes [Martin Folkes, also an antiquary and then the Society's President] and other fellows to Shoreditch, to hear Dr Denne preach Fairchild's sermon on the beauties of the vegetable world. We were entertained by Mr Whetman, the vinegar merchant, at his elegant house by Moorfields;

> a pleasant place encompassed with gardens and well stored with all sorts of curious flowers and shrubs, where we spent the day very agreeably, enjoying all the pleasures of the country in town, with the addition of philosophical company.

Whetman was without doubt Stukeley's phonetic spelling of Wittenoom, the vinegar merchant who contributed to enlarging the lecture fund: Peter Chassereau's map of Shoreditch of 1745 shows that Wittenoom occupied a substantial property on the edge of Moorfields.

In 1759 Dr Denne, whose health was failing, decided that it was time to give up the lecture, although he would remain vicar of St Leonard's until shortly before his death in 1767. For three of the next four years it was given by Stukeley, best known for his obsessive interest in Stonehenge and Druids: his garden at Grantham in Lincolnshire was laid out like a Druids' temple, at its centre an old apple tree overgrown with mistletoe. The *Dictionary of National Biography* describes him as an 'unconventional clergyman' and quotes other opinions of him as 'very fanciful' and 'a mightily conceited man'. He made frequent references to the Druids in his three Fairchild lectures. He published them in a book of sermons with a dedication to Princess Augusta (widow of Frederick Louis, Prince of Wales, who had died in 1751), but naming her 'Veleda, Archdruidess of Kew', and signing the dedication: 'Chyndonax of Mount Haemus, Druid'. In the preface he rationalised his inclusion of Druids in Christian sermons: 'Christianity is a republication of the patriarchal religion. For which reason I have not scrupled to introduce Druids before a Christian audience. They were of the patriarchal religion of Abraham . . . and have a right to assist at a Vegetable Sermon.'

He based his three Shoreditch lectures on the text from Genesis – 'Let the earth bring forth grass' – and in the first of them he paid a glowing tribute to Fairchild:

> The vegetable world was his province. It was his study, as well as his maintenance. He every day saw and admired its beauties and adored the Great Author of such admirable, such inconceivable perfection. He, like our once happy parent and progenitor, lived in a garden. But happier far: he kept possession, kept his innocence. Not content with enjoying such felicity, all his life long, he gave his substance to perpetuate that pleasure to others, which calls us together annually, as here this day.

For the next few years lecturers were engaged on a one-off basis, until in 1768 the Revd Dr Thomas Morell began a sequence that lasted until 1783, the year before he died. Morell was a literary scholar, a friend of the artist William Hogarth (who painted a portrait of him in 1762) and of the actor David Garrick. He was also an accomplished librettist, writing the words for several of Handel's oratorios, including 'See the Conquering Hero Comes'. In 1775 he was appointed chaplain to the Portsmouth garrison. His long tenure as the Fairchild lecturer suggests that the role carried considerable prestige in fashionable London.

In 1784 he was replaced by the Revd William Jones, a Fellow of the Royal Society and curate of Nayland in Suffolk, who lectured for the next four years. Each of his lectures was on a specific theme: the religious use of botanical philosophy; the nature and economy of beasts and cattle; the earth and its minerals; and finally the natural evidence of Christianity. Jones skirted the doubts that had provoked Fairchild to sponsor the lectures in the first place, stating his premise firmly in the first lecture and brooking no argument: 'We therefore take it on the authority of the text that herbs, trees, fruits and seeds are the work of God, and the present occasion requires us to consider how, and in what respects, this work is good and displays the wisdom of the Creator.'

In his final lecture, Jones dismissed scientific theories that were emerging to question that premise: 'We have been threatened in very indecent and insolent language of late years with the superior reasoning and forces of natural philosophy.' This was 25 years before the birth of Charles Darwin, whose superior reasoning would provoke a still more profound threat to established scriptural beliefs.

In 1790 the baton was passed to the Revd Samuel Ayscough, a librarian and indexer who worked in the cataloguing department of the British Museum and who had been ordained a curate in 1785. He lectured annually until his death in 1804, and in 1805 the Revd J. J. Ellis, rector of St Martin's Outwich in Threadneedle Street, set a record that is unlikely to be broken by delivering the lecture for 49 successive years until his death in 1854. None of his complete texts appears to have survived, although an extract from his 1852 lecture, published in the magazine *Sunday at Home* in 1856, suggests that the understanding of botanical processes had progressed little in the previous hundred years and that the full implications of the discoveries about plant sexuality had yet to be taken on board. Ellis again took the Genesis account of the Creation as his starting point and, according to the magazine, 'showed the analogy of nature in favour of the Christian doctrine of the resurrection of the dead, as shown by the renewal of the plant from the root, the tree from the seed and the decay and revival of all vegetable creation'.

Ellis's death brought about a crisis for the Royal Society and for the churchwardens of St Leonard's. Faced with the need to find a new lecturer for the first time in half a century, nobody was quite sure how to proceed. On 17 May 1855, 12 days before the lecture was due, 12 parishioners of St Leonard's wrote to the President, Council and Fellows of the Royal Society recommending that the Revd Hugh Hughes should be chosen. They pointed out that he had been giving weekly Sunday afternoon lectures at the church since 1836 and had afforded 'the greatest satisfaction'. Sunday lectures were a popular feature of church life in this period, and the lecturers were customarily appointed by the parishioners themselves.

There was no reply to this letter, and what happened next emerges from another letter written by the parishioners on Wednesday 30 May, the day after the lecture. As it happened, Hughes had fallen ill, so the churchwardens, 'to prevent the possibility of this bequest being lost to their parish and reverting to Cripplegate, and in pursuance of the power which we considered vested in us by the will of Mr Fairchild, appointed the Revd George Martin Braune – one of the curates of this parish – to give the lecture'.

That, however, marked only the start of the trouble, for the Royal Society, to the dismay of the parishioners, had appointed a lecturer without consulting them:

> A gentleman who gave his name as the Revd John D. Letts attended our church at 11 o'clock and stated to us that he came there for the purpose of delivering the lecture and provided a note written by Mr Weld [Charles Weld, Assistant Secretary to the Royal Society] and Sir Henry Ellis [a senior Fellow] but as Mr Letts' name was not mentioned in it, and as our vicar

> was unavoidably absent, and as Mr Letts was not known to either of our curates there present in the church or to ourselves, and as he did not produce the letters of orders or other necessary authority, under those obligations of the oath of our office and the instruction of the 50th Canon we were reluctantly compelled to object to his using the pulpit, and on our stating the same to him he withdrew.

Braune then delivered the lecture. The letter ended by giving Braune's address and demanding curtly: 'Please pay him the allowance.'

The Council of the Royal Society may have been unaware of the ecclesiastical minefield they were stepping into when they ignored the parishioners' first letter. There had been a history of controversy at St Leonard's about the appointments of preachers, especially those who gave the Sunday afternoon lectures. An incident quite similar to the Letts debacle had occurred in 1757 when the vicar, the ageing John Denne, appointed a lecturer whom the churchwardens thought had views too close to the Church of Rome. The churchwardens favoured the church's junior curate, Mr Day, and locked Dr Denne's candidate in the robing-room to prevent his ascending the pulpit. In the court case that followed the incident, evidence was given that the churchwardens then forced their favoured preacher into the pulpit, 'declaring that he should preach in spite of the vicar, the bishop, the devil or the Pope' – suggesting that they believed the senior curate to be dangerously close to the Roman faith. The churchwardens were obliged by the court to beg public pardon for their behaviour and to pay the costs of the case.

After the 1855 incident, the Royal Society decided to go for a heavy hitter as lecturer in 1856. They chose the

Rt. Revd Samuel Wilberforce, Bishop of Oxford and later of Winchester, a son of the anti-slavery campaigner William Wilberforce. He took as his text Psalm 111 – 'The works of the Lord are great . . .' – and began with praise for Thomas Fairchild:

> This lecture is his monument, and in it we may read much of his character. His daily occupation led him to be conversant with those powers of vegetable life which man may do so much to foster, and direct, and stimulate, but which it is beyond the power of all men's united force to cause, in their very lowest acting, to begin to be . . . In the special mention made in his bequest of these types of the mighty resurrection, which the annual processes of spring renewed before his eyes, there is a touching indication of the strictly practical character of his own religion. We see brought before us the pious gardener, who read the higher mysteries of grace in nature's parables.

Just three years later, the publication of Darwin's *The Origin of Species* would challenge the literal interpretation of the Genesis account of the Creation. Wilberforce never accepted Darwin's theories, and his lecture can be seen as one of the last great orations on the issue made in ignorance, accepting the biblical version without qualification. He said that the perfection of God's creations implied the existence of 'an intelligent designer', and 'their manifest oneness of purpose, their mutual interdependence, the balances and compensations, attest as plainly the unity of their Creator; whilst their infinite variety, their embodiment of matchless skill, their vastness and their littleness, all as evidently manifest the omniscience and omnipotence of Him who framed them'.

Even before Darwin's seminal work was published, his theory of natural selection was beginning to be discussed and the 1857 lecturer, the Revd Walker, was possibly the first to take it into account and to try to reconcile it with the Bible story. A report of his lecture appeared in the *Cottage Gardener* (later to become the *Journal of Horticulture*). His text was St Paul's Epistle to the Romans: 'The invisible things of God, from the creation of the world, are clearly seen, being understood by the things that are made, even his eternal power and godhead.' Walker declared that if such were the case when St Paul wrote, these invisible things could be appreciated even better since science had afforded man a more complete knowledge of God's creations: 'Science shows, and the strata of the earth reveal, that there is a progressive advance to a higher form, and all consonant with the assurance given by Revelation that there is still a higher form yet to be unfolded.'

Over the next few years, perhaps disillusioned by the unseemly spat over the 1855 lecture, some Fellows and officials of the Royal Society began to feel that it was time they tried to shed the burden of responsibility for the Fairchild Lecture. For some while, in effect since the formation of the Linnaean Society in 1788, botany had moved to the fringes of the Royal Society's agenda. Moreover, St Leonard's Church continued to be riven by doctrinal differences that made the choice of lecturer a sensitive issue in which the Society had no wish to be involved. At the end of 1872 the Society's council asked their lawyer how they could rid themselves of responsibility for the lecture. Could they not just let it slip into oblivion, claiming that the amount of the bequest was now insufficient to fulfil its purpose?

The lawyer advised that it was not as simple as that:

> We shall not get rid of the trust by failing to appoint a preacher for next Whit Tuesday . . . Had the trust been confined to Mr Fairchild's £25 I do not think we would be obliged to keep it up . . . We should be acting dishonourably if not illegally in wilfully refusing to honour the conditions of the trust.

The advice was to ask the Charity Commissioners to take over the administration of the trust and, if they refused, to seek counsel's opinion on what to do next.

In the event the commissioners did not refuse. They accepted the argument that 'the administration of the trust is not germane to the objects of the institution of the Royal Society', and in February 1873 the £100 of South Sea stock, then producing an income of £3 a year, was transferred to the Official Trustees of Charitable Funds. The appointment of lecturer once again became the responsibility of the churchwardens of St Leonard's, although they quickly ceded it to the Bishop of London.

Without the Royal Society's prestigious support, interest in the lecture seems to have dwindled. In 1895 the garden historian Alicia Amherst, the first honorary freeman of the Gardeners' Company, tried to revive the event and took a party of friends to enlarge the congregation. 'Without us there were only three or four old women there,' she recalled. The following year she tried to provoke the gardening press into publicising the lecture, but a report in the *Gardeners' Magazine* in May 1896 stated that only 30 people had attended that year's occasion.

Four years later things seemed to be looking up when the *Hackney Express and Shoreditch Observer* reported 'a large congregation' at the church to hear the Revd W. Murdoch Johnston, Vicar of St Stephen's, Twickenham. In 1902 the same paper said that 'in spite of the depressing weather a good number joined in the bright and hearty service' to hear an address by the Revd H. R. Gamble, rector of Holy Trinity, Chelsea. But by 1907 apathy had set in again, and the paper noted 'a most inadequate attendance for so notable an anniversary'.

In 1911 the Gardeners' Company, reviving after many years of inactivity, was seeking to establish a new role while suggesting a continuity of interest in the medieval guild. Diligent research – perhaps prompted by Mrs Amherst – led it to Fairchild, its illustrious eighteenth-century member whose name happily lived on in the Shoreditch lecture. Ernest Ebblewhite, the company's clerk, was asked to produce a report on the matter. After outlining the history of the lecture, his report concluded: 'I should recommend the Court to communicate with the Bishop of London and the vicar of Shoreditch Church with the object of reviving an interest in the lecture and associating this Company with the annual service at which it is delivered.' In February of that year the court approved his proposal.

Ebblewhite began a correspondence with the vicar in which a convenient time – 5 p.m. – was agreed for the lecture, to be given by the Bishop of Stepney on 6 June. Ebblewhite also promised the vicar a handsome donation of five guineas for the church's organ fund. The Worshipful Company sent invitations to all its liverymen and many local officials. The service was preceded by the laying of a garland of bay leaves on Fairchild's grave, followed by a solemn procession down Hackney Road back to the church – a tradition that survived for more than half a century.

According to *The Times*, which carried a report the following day, among those attending were the Master and Court of the Company, the Mayor of Shoreditch and several aldermen and councillors. The report went on:

> The Bishop referred to the changes in life of a butterfly or a seed, and said these were wonderful and beautiful illustrations of the resurrection of the body, but he would be an extremely bold man who went further than that in the present day and tried to prove, as Fairchild would apparently have had them do, the certainty of the resurrection from the changes of animal and vegetable life. God's truth remained unchanged but the evidence on which that truth might be supposed to rest varied from time to time.

Certainty was less in style in the early twentieth century than it had been in the early eighteenth. The bishop added that Fairchild's book about what grows best in London 'might with advantage be brought up to date', but no publisher acted on the hint.

The Times's interest in the lecture was not sustained. The following year it limited itself to an announcement that the lecture was to take place on 28 May and a report that it had duly done so, with the Archdeacon of London as the lecturer. Other reports appeared in the local and gardening press from time to time, and it is clear from them that the prevailing theme of the lectures in the twentieth century was an attempt to reconcile Fairchild's now somewhat archaic beliefs, as reflected in the lecture subjects, with modern science. Although it was now accepted that the biblical version of the Creation could not be regarded as literal fact, the lecturers insisted that it did not affect the fundamentals of Christian belief.

One of the few who seemed aware of Fairchild's role in the development of hybridisation was the Revd E. W. Barnes, Canon of Westminster, who delivered the lecture in 1923. He asked what it was that made possible the varieties of flowers and fruits that gardeners could today enjoy:

> From the dog rose, [God] has created the splendid multitude of garden roses which we enjoy. From the crab apple has come the Ribston pippin . . . God's secrets are still hidden, but we have advanced a step on the long, perhaps interminable road which man must travel before he understands the mechanism of the universe of which he is a part.

He concluded that it was Fairchild's faith that had inspired the man of science to find laws behind the apparent confusion and an order in the universe that hinted at purpose: 'In the future progress of mankind upon the earth, religion and science must go hand in hand, no longer enemies but friends.'

The *Hackney Gazette* gave fairly regular reports on the lecture in the first half of the twentieth century, with some intriguing details. In 1932 the choir was augmented by the manager and cashiers from the Shoreditch branch of Barclays Bank, and members of the Gardeners' Company wore buttonholes of carnations and roses. The lecturer was the Revd Maurice Relton, Professor of Theology at the University of London, one of several lecturers during this period who took issue with Fairchild's premise that the wonders of nature proved the certainty of resurrection: indeed, later discoveries about evolution tended to suggest the opposite. 'The only possible certainty which we could have of resurrection was that which was given in the Christian message.'

In 1935 the Revd Prebendary Tom Wellard, chaplain to the Gardeners' Company, reverted to the theme of earlier preachers that the sheer beauty of God's creations proved beyond doubt that they were designed by a supreme being, and that Fairchild as a gardener was in the perfect position to appreciate this:

> A mere blind evolution could never conceive or paint the flowers. The philosophy of a soulless materialism was refuted absolutely by the pageantry of sea and sky . . . Why were heavens blue and not brown or grey? Why did the waves break in sapphire and white upon the yellow sands?

He concluded by observing that 'devotion to his garden plot had kept many a man's feet from straying in the paths of vice'.

The lecture continued through the two world wars, although from 1940 the procession was suspended. In 1944 the Revd Max Petitpierre was one of the few lecturers to take a botanical approach to the subject:

> No matter what method of propagation was adopted, the nuclear cells in the protoplasm succeeded in reproducing the entire plant in its normal and recognisable habit. That attainment of organic completeness was more characteristic of the vegetable than of the animal world . . . The resurrection of the dead was analogous. Man, human life as they knew it, was compounded of two things – body and spirit – and it was just the combination of those two which was the characteristic habit of growth of the human species.

For three years from 1945 the lecture took place in other churches, because St Leonard's had been badly damaged by a flying bomb in August 1944. In 1948 it went back to St Leonard's, and the procession was resumed. In 1952 the Bishop of Woolwich gave the lecture and made headlines by criticising the BBC's 'pagan' commentary on the funeral of George VI the previous year. The radio commentator had said that the king lay in St George's Chapel 'guarded for ever by the great tower of Windsor'. Taking up Fairchild's theme of resurrection, the bishop commented: 'The death which ends a man's life in the earth is real death. Sentimental people speak of "passing on", but you do not pass on: you stop. You stop dead. But somehow out of that death comes a new life, a higher life.'

In the postwar years the *Gazette*'s reports of the wreath-laying and lecture became more spasmodic and tended to concentrate on the disruption caused by the procession in the evening rush-hour. In 1950, for the first time, a note of unambiguous scepticism was introduced when the paper reported: 'As usual, there was almost a complete lack of interest taken in the service.' The writer added that some months earlier the Shoreditch borough librarian, C. M. Jackson, had sought to stir up interest by giving a talk on the lecture on the BBC Home Service – an effort that 'merited greater success'. A rather more effective tribute to Fairchild's memory was made in September of that year when the Minister of Health, Aneurin Bevan, officially opened Fairchild House, the first block of flats completed on the new Pitfield Estate, a few hundred yards from the former nursery.

On 2 June 1958, under the headline 'Gardener Thomas is remembered', the *Gazette* reported: 'Workers going home in Hackney Road, Shoreditch, on Wednesday night stopped to watch a procession of gowned men, all carrying a small green bunch of herbs and led by a man holding aloft a shining spade . . . flanked by policemen and followed by small boys.' In 1966 ('Gardeners keep up tradition') the ceremony took place 'beneath a blazing sun', as it did the following year, when the *Gazette* carried a picture of 'a solemn moment as members of the Most Worshipful Company of Gardeners and the vicar of Shoreditch, Revd R. I. Thomson, stand in silence at the laying of the wreath on Fairchild's tomb'. At that time the land at whose edge the tomb stands was used as a public tennis court, now gone.

As it happened, that was the last time for seven years that the lecture was held in St Leonard's Church. In 1968 it fell victim to the growing climate of ecumenicism in the Church of England and the Catholic Church. It was the year when the Archbishop of Canterbury was invited to preach at the Catholic Westminster Cathedral, and he reciprocated by inviting Cardinal Heenan, head of the Roman Catholic Church in England, to preach at Westminster Abbey. Falling in with the progressive mood, the Gardeners' Company invited the Catholic Bishop Butler, former Abbot of Downside, to give the lecture.

Not all members of the Church of England went along with the new ecumenicism, and it is doubtful whether Fairchild, living as he did at a time when Catholics were regarded with grave suspicion, would have approved the choice. Richard Thomson, the vicar of St Leonard's, was persuaded by his churchwardens not to allow a Catholic to preach in their church. The Gardeners' Company refused to disinvite the bishop and had to find an alternative venue at short notice. That year's service was switched to St Michael's in Cornhill.

From 1969 to 1975 the lecture was given in Fairchild's

reserve choice, St Giles, Cripplegate. A report of the 1971 lecture showed how liberal the speakers' interpretation of Fairchild's themes had become, as they bent them to drag in contemporary references in the modern manner of BBC Radio's *Thought for the Day*. According to the *City Press*, Canon Clifford Earwaker, Prebendary of Chichester, gave an address in which 'man's first journey to the moon and the latest scientific concept of the universe were compared with a butterfly breaking from its chrysalis and the dividing of the atom'. He even managed to incorporate a tribute to the broadcaster Richard Dimbleby, who had recently died.

In 1976 the lecture went home to Shoreditch but remained there for only five years. The late 1970s were a time of political polarisation in Britain, especially in the depressed boroughs of Inner London. Margaret Thatcher's Conservatives gained power nationally in 1979, but in local government the hard left was taking over the ruling Labour groups. The flummery associated with the lecture and the wreath-laying were not at all to the taste of the champions of the humble who ran Hackney Council. The mayor and local officials declined to attend, and passers-by, witnessing the procession to Fairchild's grave, saw only an overspill of the City's arcane habits and fancy dress, as well as a traffic snarl-up. In 1981 the lecture moved back to St Giles – where for the time being it remains – and the wreath-laying ceremony was suspended. Fairchild would have been gratified that his creation has survived more than two and a half centuries, with every prospect for its continuance – gratified and, as a supreme realist, more than a little surprised.

CHAPTER NINE

Floribunda

It is impossible to doubt that there are new species produced by hybrid generation . . . Thence it appears to follow that the many species of plants, in the same genus in the beginning, could not have been otherwise than one plant, and have arisen from this hybrid generation.

LINNAEUS, *DISQUISITION ON THE SEX OF PLANTS*, 1760

Stephen Bacon might well have proved a worthy successor to Thomas Fairchild. He took the job seriously. He had joined the Society of Gardeners – though not the Worshipful Company – and had contributed to its *Catalogus Plantarum*. Mark Catesby, whose work in the American colonies had by now brought him renown in botanical circles, chose to stay on at the nursery for four years after Fairchild's death even though, with his reputation, he could easily have found a position elsewhere. This suggests that Catesby had been impressed with the manner in which Bacon had stepped into Fairchild's shoes during his final illness, and had every confidence in his ability to carry on the pioneering work at Hoxton with the same success. Catesby's presence helped sustain the nursery's hard-won prestige, and in March 1730 the third Earl of Burlington – the aristocrat who had snubbed Bradley in the Netherlands 16 years earlier – went to Hoxton to see Catesby and perhaps choose some plants for his garden at Chiswick House, which had just been laid out. Bradley continued to use the nursery for his experiments. In *The Fruit Garden Display'd* he records that he propagated several small black marmalade figs, or honey figs, 'in Mr Bacon's garden at Hoxton'.

It cannot have helped the transfer of responsibility that, only weeks after Fairchild's death, Bacon had to endure another bereavement when his younger brother John died at the age of only 17. He was buried alongside Fairchild, in the poor people's graveyard by St Leonard's, on 2 December 1729. Yet for Stephen life went on; that unhappy winter eventually gave way to spring and summer, and in the autumn of 1730 he married Mary Baynham at St Giles, Cripplegate. In June 1732 they had a son and, in a touching tribute to his uncle and benefactor, they christened him Fairchild Bacon, following the example set by their neighbour Benjamin Whitmill 11 years earlier. Their marriage, though, was destined to be a short one. In February 1734 Stephen died, at the age of 25, and Thomas Fairchild's grave gained a third occupant.

With no obvious heir among Stephen's relatives, and her son less than two years old, the young Mary Bacon arranged the sale of the nursery to John Sampson, who had worked for Fairchild for some years and had been left £5 in his will. But his tenure did not last long either, for in April 1739, he, too, died, while his wife Anne was pregnant with their fifth child – or their ninth, if you include the four who had died soon after birth. All these deaths seem early by today's standards, but Sampson had attained the age of 39, and three-quarters of people born in the eighteenth century died before they were 40 – more than half as infants. Anne may have tried to keep the nursery going for a while. She paid the poor rate for 1740, but later that year an announcement of the sale of the nursery's stock and effects was made in the newspapers.

Although his nursery did not long survive him, Thomas

Fairchild's name lived on through the memorial lectures. Before long, though, it became just a name, scarcely associated with an actual memory of the man. In 1760 a pirated edition of *The City Gardener* appeared under the title *The London Gardener*, almost word for word the same as the original but with no attribution to Fairchild. This cannot have been for reasons of copyright, since under the copyright statute of 1710 the protected period was limited to 28 years, so it would have expired in 1750. Fairchild's name was presumably omitted because the publisher calculated that it would not help to sell the book. Yet his real contribution to horticulture has survived far longer, and the gardens we enjoy today owe an enormous debt to the techniques he pioneered.

After Fairchild's death, Grew's and Camerarius's theory of plant sexuality, which Fairchild had been the first to put into practice, gradually came to convince naturalists, with fewer and fewer diehards coming forward to contest it, despite its being in conflict with the biblical account of the Creation. In 1749 Johann Gottlieb Gleditsch, director of the Berlin Botanic Garden, wrote: 'Things which most physicians formerly regarded as ridiculous and imaginary are proved today by the most simple experiments, and with so much evidence that there no longer remains the least place for any objections being made against the theory [of sexual reproduction in plants] or for all the jests that could accompany it.' (This was 17 years after the publication of 'The Natural History of the Arbor Vitae, or the Tree of Life', which had encouraged a spate of similar satires.) He added that only a small number of people now doubted the theory, and that 'their arguments do not appear to merit any response'.

None the less, and rather late in the day, the Imperial Academy of Sciences at St Petersburg in 1759 offered a prize for a convincing proof that plants had a sex life. The following year it was awarded to Linnaeus, by then the doyen of European botanists, whose prize-winning essay was essentially a cogent roundup of the research of the previous hundred years, presented in such a way as to make the premise irrefutable. He described his own experiments in the hand-pollination of plants that had been prevented from pollinating naturally. For the most part he was working with flowers of the same variety, but he does describe a few experiments in hybridisation, including one between two species of the genus tragopogon: *Tragopogon pratensis*, commonly known as goat's beard, and *Tragopogon porrifolius*, usually grown for its root and eaten as salsify.

'I do not know whether any other experiment would show generation more certainly,' Linnaeus wrote, noting that it provided an opportunity for botanists to try to create new kinds of vegetables. It suggested to him that the various existing types of brassicas and lettuces had been formed by spontaneous hybridisation, rather than, as had formerly been supposed, because of differences in the type of soil they grew in. Philip Miller had made the same suggestion in his *Gardeners' Dictionary* some years earlier.

The Linnaean system of plant classification, accepted in Britain from about 1760, relies to a large extent on his recognition of the method of sexual reproduction, and for this reason some botanists resisted its introduction for some time. Keith Thomas, in his book *Man and the Natural World*, writes of 'prudish objections' to the Linnaean system's supposedly 'licentious' character and adds: 'Botany seemed a doubtful recreation for young ladies when it involved so close a scrutiny of the "private parts" of wild flowers.' (Yet despite – or perhaps because of – this risqué element, the study of natural history became an increasingly popular pastime among women of the leisured classes in the eighteenth century.)

One of the most ferocious critics of Linnaeus and his theories was Charles Alston, the King's Botanist in Scotland and Professor of Botany at the University of Edinburgh, who believed that the concept of sex in plants derived from too close an analogy between animals and vegetable matter. 'Analogy alone proves nothing,' he wrote in a pamphlet in 1754. 'It would not be worthwhile to argue against the sexes of plants unless it had given occasion to the specious contrivance of a system or method of plants, named sexual, which of all others is the most intricate, involved and unnatural.' Man's role, he believed, was to receive and make use of the species handed down by God, not to engage in the 'wanton and useless destruction of God's works'.

Others objected to the Linnaean system because it classified plants essentially by their physical attributes rather than by the medicinal use to which they could be put – or by popular lore and superstitions about their powers and origins – which had been the basis of earlier classifications. Linnaeus was rigorous about this, declaring: 'If a genus known for a long time past, and familiar even to ordinary people, should bear an absolutely erroneous name, it should be expunged.' In researching his work on plant naming he had come to Britain in 1736 and visited the Oxford Botanic Garden and the Chelsea Physic Garden, where he was shown round by Philip Miller, who, after his initial suspicions – based largely, it seemed, on his innate stubbornness – was to become a leading advocate in England for the Linnaean system. A persistent story that Linnaeus had earlier corresponded with Fairchild cannot be confirmed: the Linnaean Society in London has no trace of any such correspondence, and Linnaeus would have been only 22 when Fairchild died in 1729.

The first scientist to experiment methodically and extensively with hybridisation was Joseph Gottlieb Kölreuter, a professor of natural history and curator of the Karlsruhe Botanic Garden in Germany from 1764 to 1806. In 1760, after what he described as 'many experiments instituted in vain with many kinds of plants', he achieved his first successful deliberate hybrid, of two species of nicotiana (tobacco plant). Like Fairchild's mule, it turned out to be sterile. Over the next six years Kölreuter created at least sixty-five other hybrids. He was exhaustive and meticulous in his methods, inserting different amounts of pollen from the alien flower, and mixing it with pollen from the receptor in different proportions, to see what effect that had on the hybrid's colour and other characteristics. He even mixed the pollen with different kinds of vegetable oils to determine whether this would increase or reduce the chances of successful hybridisation. And he made full use of the rapidly improving microscope technology to examine pollen and seeds in close detail, so that he could understand the mechanics of cross-fertilisation as well as its consequences.

He was among the first to recognise the importance of insects as pollinators, noting that they perform 'this uncommonly great service' by carrying pollen between flowers that are not necessarily of the same species: 'Almost all flowers . . . carry something with them that is agreeable to insects, and one will not easily find one of them with which they are not to be found in quantity.'

Kölreuter noted that hybrids could normally be created only out of species that were related to one another quite closely: it would be 200 years before genetic modification allowed hybridisation across a wider range. From this observation he formulated a novel theory about why comparatively few random crosses occurred naturally. It was, he said, because nature had 'enclosed within the boundaries of a certain region only such plants as, in regard to structure, have the least resemblance amongst themselves and which, consequently, are also least qualified to cause a confusion among themselves'. It followed that the creation of botanic gardens, where plants from all regions and climates were artificially brought together, could theoretically give rise to a number of spontaneous hybrids. Further experiments showed that this seldom happened, because of another of nature's built-in safeguards: if flowers are pollinated at the same time by pollen from their own and another species, they normally choose to breed from their own kind.

Kölreuter's most far-reaching discovery in terms of today's horticulture was to do with double flowers. He noted that if you cross a double-flowered variety with a single-flowered, the result is invariably a double-flowering plant – foreshadowing Mendel's work on the dominant gene some years later. Hybrids also tend to be more vigorous in growth and stay in bloom for longer than plants of a single species, probably because, being often barren or of low fertility, they do not expend their energy producing seed. Charles Darwin confirmed this nearly a hundred years later in *The Origin of Species*, where he wrote:

> Both with plants and animals, there is the clearest evidence that a cross between individuals of the same species, which differ to a certain extent, gives vigour

> and fertility to the offspring; and that close interbreeding continued during several generations between the nearest relations, if these be kept under the same conditions of life, almost always leads to decreased size, weakness and sterility.

The identification of hybrid vigour was the first pointer to using hybridisation to improve the quality of flowers and vegetables available to gardeners and not merely, as Fairchild had done, to increase the range. Yet in the decades following Fairchild's death, there was scarcely any commercial exploitation of the techniques he pioneered. The emphasis was still on novelty, on introducing entirely new species of plants from overseas rather than breeding variations on existing ones.

Among foreign plants popularised in Britain in the eighteenth century were sweet peas, chrysanthemums, dahlias, fuchsias, auriculas and nasturtiums. In time all these flowers would provide rich veins of material for hybridisers – but not yet. Gardening was still essentially a branch of collecting, of acquiring rare and spectacular plants chiefly for the pleasure of possession. The idea that you could design and manufacture new flowers with attributes of your own choosing would change it into a more creative pastime, a participatory instead of a spectator sport.

Even without that critical element, though, it was in the eighteenth century that gardening became a permanent part of the national consciousness, with a greatly broadened appeal. Keith Thomas calculates that while 131 new species of tree had been introduced in the seventeenth century, the number in the eighteenth was 445, including the Lombardy poplar, spruce and larch. By 1760 England boasted 10 garden designers, 150 noblemen's gardeners, 400 gentlemen's gardeners, 100 nurserymen, 150 florists (specialist flower growers), 20 botanists and 200 market gardeners. And the status of the gardener was rising rapidly. In 1764 a visitor to Thomas Mawe, head gardener to the Duke of Leeds, found him 'so bepowdered and so bedaubed with gold lace' that he took him for the duke himself. Gardening books were proliferating, too. John Abercrombie's *Every Man His Own Gardener* (1767) had gone into 16 editions by 1780. *The Gardener's Pocket Journal*, first published in 1789, sold 2,000 copies a year. In 1787 came the first gardening magazine (if you discount Bradley's efforts), the *Botanical Magazine or Flower Garden Displayed*.

Two developments in that pivotal century would eventually form a springboard for the growth of hybrids. One was the establishment in 1759 of the Royal Botanic Garden at Kew under the auspices of Princess Augusta – although in its early years it specialised in new plants introduced from overseas rather than on botanical experimentation. The second was the growth of the annual flower show. As early as the sixteenth century, groups of refugees from France and the Netherlands had formed societies to promote flowers that they had been accustomed to cultivating at home, and organised competitions to see who could grow the best. Over the next 200 years more societies for individual flowers were set up by enthusiasts, but these came chiefly from the new breed of working-class gardeners, striving to keep something of their rural origins even though they were now crammed into the crowded industrial towns and cities where they had migrated, seeking work. These societies would eventually form the seedbed of the insatiable demand for new flowers that would allow hybridisation to serve a mass market.

Eight kinds of flowers were being grown for exhibition by 1790. These were hyacinths, tulips, ranunculus, anemones, auriculas, carnations, pinks and polyanthus – the so-called florists' or mechanics' flowers or, as David Hessayon called them more bluntly in his *Armchair Book of the Garden*, 'working-class plants'. The first *Florist's Directory* was published in 1792, written by James Maddock, a nurseryman who stocked 800 varieties of ranunculus. Due in part to its newly acquired working-class connotation, the ranunculus was henceforth shunned by the wealthy owners of estates, who preferred the larger and showier plants that were being brought from the east by the plant-hunters, as providing clear evidence of their fortune and status. In 1804 these upper-class gardeners formed the exclusive Horticultural Society (it would become 'Royal' later in the century), which until the twentieth century exemplified the divide between the two kinds of gardener. The Society held its first flower show at Chiswick in 1827.

The ranunculus, a relative of the buttercup, had been the fashionable flower of the mid-eighteenth century. Clusius had introduced it into western Europe from Turkey more than 200 years earlier. Fairchild knew it well, and it occurs several times in the list of plants he supplied to Bradley's journal. The first month he reports a ranunculus in flower is March, presumably under glass. In April the plantain-leaved and double white mountain varieties appear, in May 'ranunculus of the Persian kind' and in June 'fine ranunculus', as well as the double white mountain variety making a second appearance. Members of the industrial flower clubs prized these double flowers most. As a champion of the dispossessed, Fairchild would have been delighted that his pioneering work had helped to broaden the class spectrum of gardening enthusiasts. Only in the late twentieth

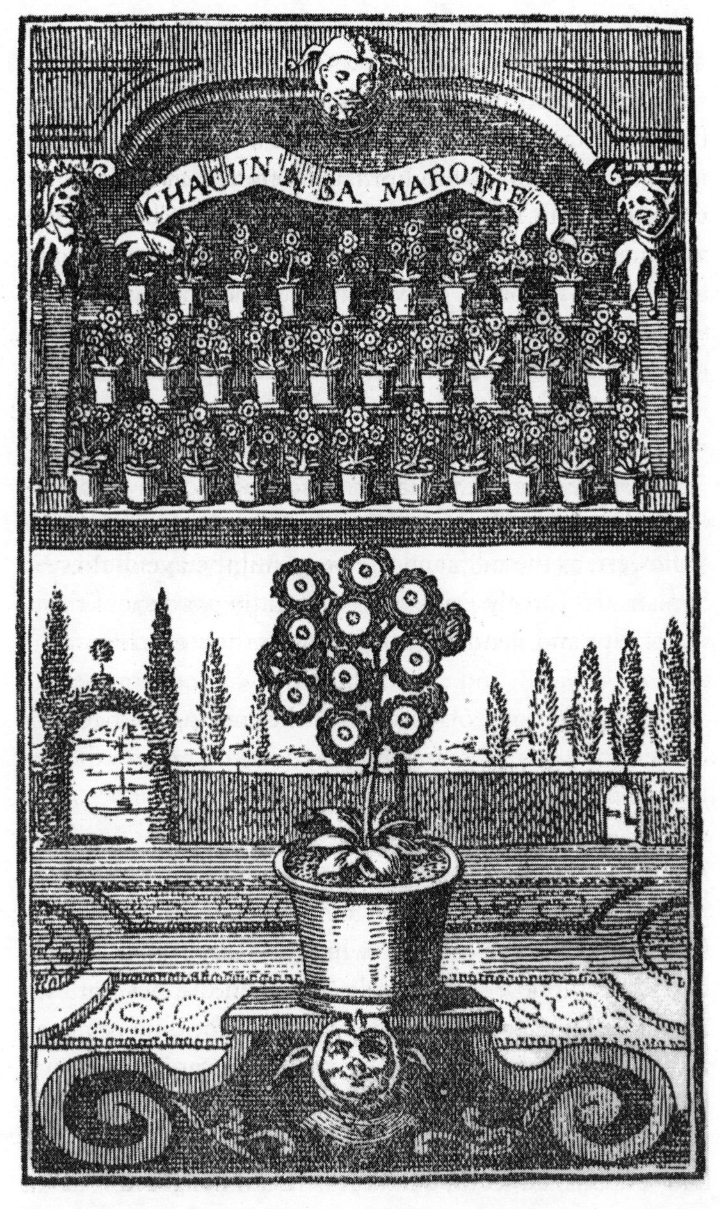

Auriculas illustrated in a French book of 1738. The motto means: 'Everyone has his fancy' (*Royal Horticultural Society, Lindley Library*)

century did the ranunculus return to high fashion as a desirable cut flower for the home.

Fairchild also grew auriculas, or bears' ears, which had been in Britain since Tudor times, and he lists them as being in flower from January to April. Later they became a particular favourite of cotton weavers in Lancashire: one of their advantages is that they do well in pots, so they can be grown by people without a conventional garden. A book called *Flora Domestica*, published in the early nineteenth century, leaves no doubt about the flower's social implications:

> The auricula is to be found in the highest perfection in the gardens of the manufacturing class, who bestow much time and attention on this and a few other flowers, as the tulip and the pink. A fine stage of these plants is scarcely ever to be seen in the gardens of the nobility and gentry, who depend on the exertions of hired servants, and cannot therefore compete in these nicer operations of gardening with those who tend their flowers themselves, and watch over their progress with paternal solicitude.

Another group of cotton weavers, this time from Paisley in Scotland, specialised in growing pinks and developed the laced or Paisley pink, whose white petal, delicately fringed in pink, is supposed to have derived from the lacy patterns they were engaged in weaving at the factories. They bred these pinks from seed, not by hybridisation but by carefully segregating those with the desired flower pattern so that they did not breed with the plainer flowers – a long-established technique that is in a sense the reverse of hybridisation, but is equally manipulative of 'nature' and shows that the principles behind plant breeding were gradually beginning to be understood. Other specialities of the northern industrial areas were pansies in Derbyshire and polyanthus on both sides of the Pennines.

Many of today's most widely exhibited flowers are missing from that 1790 list. One is the fuchsia, whose bumpy career as a flower of fashion illustrates the mixed fortunes that can befall a single plant, until hybridisation eventually established its place in the affections of present-day gardeners. The shrub, with its masses of pendulous red flowers, was discovered in South America in the late seventeenth century and named after the pioneering sixteenth-century German botanist Leonhard Fuchs – as a tribute to his memory and not because he had anything to do with its discovery. Fairchild did not grow it, and the flower is not recorded in Britain until the mid-eighteenth century, when, in the manner of tulips and hyacinths a hundred years earlier, it became sought after and consequently expensive.

A romantic story surrounds the introduction of *Fuchsia magellinaca* by the Hammersmith nurseryman James Lee in the 1790s. In *Garden Shrubs and Their Histories*, Alice Coats casts some doubt on its authenticity but, whether true or not, it is symptomatic of the mystique surrounding rare and unusual plants at the time. A customer at the Vineyard Nursery, owned by Lee and John Kennedy, is said to have told Lee that he had seen a fine example of a new kind of fuchsia in the window of a hovel in Wapping, by the docks in East London. When Lee went to see it, the woman of the house said it had been given her by her husband, a seaman, who had acquired it on his travels. At first she refused all Lee's offers to buy the plant but eventually accepted eight guineas. Getting it back to the nursery, Lee propagated about 300 new plants from it and sold them for a guinea each. 'Chariots flew to the gates of old Lee's nursery grounds', according to a newspaper account, and the wily nurseryman made a handsome profit on his excursion to Docklands.

Yet despite their popularity, the only fuchsias then available were species plants gathered from overseas and propagated by cuttings. Not until the 1830s were the first hybrids attempted, usually crossing *Fuchsia magellinaca* with the newly introduced *Fuchsia fulgens*. In 1842 Lucombe and Pince, nurserymen of Exeter, introduced the hybrid *Fuchsia exoniensis* (now called *corallina*), and in the same year Thomas Cripps of Tunbridge Wells produced a variety with white sepals, which he also sold for a guinea.

By 1848 there were 520 species and varieties available. The first double flower was produced in 1850 and the first three-coloured variety in the 1870s. James Veitch, the celebrated Chelsea nurseryman, was one of several enthusiasts who went to work on developing the breed: by 1880 the number of varieties had nearly tripled, and an estimated 10,000 fuchsia plants were sold every day at Covent Garden flower market. Giant specimens up to ten feet high were proudly exhibited at flower shows.

In the early years of the twentieth century the fashion for fuchsias waned as the hereditary estates began to disappear, taking with them the glasshouses and conservatories and the staff needed to sustain them. Interest revived towards the middle of the century, with the formation of the British Fuchsia Society in 1939 and the breeding of hardy varieties that could be left outdoors in winter: in Ireland and milder parts of England they have become naturalised. Lawrence Johnston introduced an area devoted to fuchsias into his renowned garden at Hidcote in Gloucestershire. Today, they are part of the stock in trade of garden centres and nurseries and cost a lot less than the equivalent of an eighteenth-century guinea.

Veitch's other main contribution to the advance of hybridising was in the exotic, not to say erotic, world of the orchid. The steamy sensuality of this spectacular flower was subconsciously recognised by the Greeks, whose name for it derived from 'orchis', their word for testicles: this is because some species have pairs of root tubers of a testicular shape. Yet despite giving it that name, the Greeks were probably unaware of its highly unusual reproductive arrangements, with its male and female attributes fused in a single column, equipped with sophisticated lures for bees and other pollen-transferring insects. (Some species of orchid have parts that so resemble female insects that the males attempt copulation, get covered with pollen and then spread it to another flower similarly disguised – frustrating for the insect, but it gets the job done.)

All that would have intrigued Fairchild, but the exotic orchids from the tropics did not reach England until 1731, two years after his death – brought, as it happened, by Peter Collinson, who had visited his nursery some years before. It took almost a hundred years for the flowers to become available to the general trade, mainly through the initiative of Conrad Loddiges, a Dutchman who ran a nursery at Hackney, not too far from Hoxton. After he died in 1826, his son George inherited the nursery and his 1839 catalogue listed 1,600 orchids, none of them a hybrid. By this time orchids were attracting a group of dedicated enthusiasts, including the sixth Duke of Devonshire, who raised them on his estate at Chatsworth, Derbyshire, in the innovative glasshouses built for him by Joseph Paxton. James Bateman, founder of the magnificent garden at

Biddulph Grange in Staffordshire in 1842, was another early orchid fancier.

Charles Darwin did some pioneering studies on orchid reproduction, and they served to buttress his theories on evolution: 'Why do orchids have so many perfect contrivances for their fertilisation?' he wondered. But it was one of Veitch's employees, John Dominy, who in 1854 produced the first man-made orchid hybrid, named after him as *Calanthe* x *dominii*. Orchids hybridise more readily than some other plants, producing fertile flowers even from matches between distant relatives. The great variety and complexity of their colouring has led to thousands of orchid hybrids being produced – one estimate puts the total at 100,000. Royal patronage boosted their popularity still further: Queen Victoria was fond of them and so were her cousins the Romanovs, emperors of Russia. Soon, collecting orchids became a universal passion among those who could afford the high prices.

As the nineteenth century progressed, hybridisation came to be accepted as the most efficient and productive method of creating new varieties of most kinds of flowers. John Seden, Dominy's successor at Veitch's, did pioneering work on gloxinia, begonia, hemerocallis and many other plants, including new strains of soft and hard fruit. Some hybrids were so clearly more desirable than their parents that the original species virtually dropped out of cultivation, and the hybrids came to be accepted as the standard forms. This has happened with clematis, dahlias, astilbes, delphiniums, phlox, gladioli and many other garden favourites.

In addition to their pioneering work with orchids, the Loddiges are credited with the introduction in the late eighteenth century of another plant that was to provide scope for hybridisers: *Rhododendron ponticum*, the run-of-the-mill mauve shrub, initially brought in from Gibraltar, that is now rampant in some woodland areas of Britain. The first member of the rhododendron family brought to Britain was the small R. *hirsutum*, or alpine rose, which came from Switzerland in the late seventeenth century. It was hybridised with other species from America and Europe (which boasts only four, including *ponticum*) to produce flowers in a limited range of pinks and pale purples.

In the early nineteenth century the palette of colours expanded with the importation of the bright red R. *arboreum* from India. Then came R. *catawbiense*, or mountain rosebay, from North Carolina, whose hardiness made it popular with the hybridisers. The best early hybrid from those two species was *altaclarense*, raised by the Earl of Caernarvon at Highclere in Hampshire in 1826. John Lindley wrote of it in the 1831 *Botanical Register*: 'It not only shows how great the power of man is over nature, but holds out to us a prospect of the most gratifying kind in regard to the future gayness of our gardens.'

All this whetted the appetites of collectors for the ever more exotic rhododendrons and azaleas brought back from China and the Himalayas as the century progressed. Here was an area of horticulture that the landed gentry could still claim as their own, now that the currency of florists' flowers had been devalued by the enthusiasm of the labouring classes. It cost a great deal to fund plant-hunting expeditions by such as Joseph Hooker, Robert Fortune, Ernest 'Chinese' Wilson and Frank Kingdon-Ward, and you needed extensive grounds to display the spoils to their best advantage. Aristocrats such as the Rothschilds at Exbury, Hampshire, the Williamses at Caerhays, Cornwall, and Sir Edward Loder of Leonardslee in Sussex worked on the material to produce spectacular hybrids in ever more dazzling colour combinations.

The rose reaped the most dramatic benefits from hybridisation. It had been a prominent garden flower for some 2,000 years, and it is likely that the Chinese had been hybridising it for centuries. Shakespeare mentioned it repeatedly, and in the mid-seventeenth century John Tradescant had 32 species in his Lambeth garden. Nicholas Culpeper, in his *Complete Herbal* of 1652, recommends roses for a variety of medicinal purposes. Red ones strengthen the stomach, control vomiting and relieve tickling coughs and consumption; white ones are good for inflamed eyes, while damask roses form the basis of a syrup that makes an excellent purge for children.

Fairchild, though, seems not to have had much time for them. In *The City Gardener* he made a list of 11 flowering shrubs that perform well in London and included only one rose, 'the province rose, white and red', which we know as *Rosa gallica*. He commented: 'No other sort of rose will stand in the City gardens since the use of sea-coal; though I am informed that they grew very well in London when the Londoners burnt wood.' Certainly, London clay is ideal for them, as millions of gardeners in the Victorian suburbs would later discover.

In the *Catalogus Plantarum*, published by the London Society of Gardeners in the year following Fairchild's death, 43 roses are named. The range increased significantly over the next century, and in 1836 Thomas Rivers, a leading rose grower, listed nearly 600 varieties, many of them hybrids. For a time hybrid perpetuals captured the Victorian market – they dominated the first Grand National Rose Show in 1858 – but they were ungainly, especially in smaller gardens, and in 1867 came the pink double-flowering La France, a cross between a hybrid perpetual and the more delicate tea rose. Its neat shape and pretty flower head made it an immediate success in Europe and America and encouraged numerous similar crosses.

Hybrid teas were not formally recognised as a class until 1880, largely through the efforts of Henry Bennett, a Wiltshire farmer who was one of the first to apply to roses the systematic breeding principles that he used with his livestock. Bennett bred more than 30 new varieties of rose in the last 10 years of his life, and the wide range of beautiful and vigorous hybrid teas would change the look of our gardens permanently.

In the early years of the twentieth century hybrid musk roses came on to the scene, and in modern times David Austin has further developed the principles of hybridising with his English roses, crosses between the old types such as *gallica* and the damask rose. After careful selection of the resulting seedlings, the new roses put into production have the vigour and repeat flowering characteristics of hybrid teas combined with the elegance and usually the scent of the older sorts. Fairchild would not have recognised them, but he would probably have been proud to grow and sell them.

One notable omission from the 1790 list of exhibition flowers is the sweet pea, although it had been grown in Britain since 1699, when Dr Robert Uvedale of Enfield successfully planted the seed sent to him by the Sicilian monk. Fairchild sang its praises 23 years later in *The City Gardener*:

> The sweet-scented pea makes a beautiful plant, having spikes of flowers of a red and blue colour. [He is

> describing a single purplish colour, because pure blue sweet peas were not introduced until much later.] The scent is somewhat like honey and a little tending to the orange-flower smell. These blossom a long time.

But by the end of the eighteenth century there were still only five colours in cultivation, because the sweet pea is self-fertile and is thus scarcely ever pollinated by another flower. Only by human-assisted hybridisation could a start be made on amassing the huge number of varieties available today, and this would not happen until the mid-nineteenth century.

In 1860 Colonel Trevor Clarke of Daventry, trying to produce a pure blue sweet pea, created a white one edged in blue. In 1870 Henry Eckford, known as the father of the modern sweet pea, got to work with his pollen brush to introduce more than 100 new crosses. At the Bicentenary Sweet Pea Exhibition in 1900, 264 varieties were shown, nearly all of them hybrids and nearly half created by Eckford. A year later the still popular Spencer varieties, with their characteristic wavy petals, were developed by Silas Cole, gardener at the Earl Spencer's house at Althorp, Northamptonshire, who had been experimenting by crossing some of Eckford's hybrids. This enormous variety, coupled with the flower's superb scent and prolific habit, made the sweet pea, in the words of Alice Coats, 'the Edwardian flower *par excellence*'.

The nineteenth-century geneticist Gregor Johann Mendel had earlier used table peas for the critical experiments that allowed him to understand precisely how the characteristics of plants, and by extension animals, were passed from one generation to the next. His work provides the bridge between the eighteenth-century botanists – including Fairchild – who identified plant sexuality and the twentieth-century geneticists who discovered how to transfer genetic characteristics between species. A farmer's son, born in Austria in 1822, Mendel entered a monastery in Brno (now in the Czech Republic) in 1843, where he taught himself the rudiments of science. After being ordained a priest in 1847, he was sent by his abbot in 1851 to the University of Vienna and studied natural sciences.

In 1854 he returned to Brno but kept up his scientific studies. In 1862 he joined the newly formed Natural Science Society in the city, and made sure that he read the latest learned works, including those of Darwin. He took an interest in the emerging evolutionary theories as well as in plant hybridisation, and saw the link between the two. He also recognised that hybridisation would have to be carried out on a more methodical basis than hitherto if the precise role of heredity in species was to be established incontrovertibly.

For the want of any evidence to the contrary, it had until then been assumed that when you crossed, say, a red flower with a blue variety, the reproductive material from the two parents would mix in different proportions and produce flowers in varying shades of mauve. What puzzled Mendel was that if you then crossed two of the mauve flowers, not all their offspring would inherit that colour, but some would revert to red or blue.

He worked out a theory that would explain this. Each of a plant's attributes is determined by a pair of what we now call genes, one inherited from each parent. Thus, a mauve-flowered hybrid may have inherited a red and a blue colour gene, only one of which will be passed on to its offspring. In some cases, therefore, a hybrid from two mauve parents will inherit two red genes and produce a red flower. That explains why seeds of today's F1 hybrids, those taken from two pure-bred parents, produce nearly uniform plants; but when you try to produce a second generation from the seeds of the hybrid there will be considerable variation.

Mendel confirmed this theory by a series of experiments with peas. Noticing that the peas grown in the monastery garden all varied in minor ways – height of plant, colour of seed, shape of pod – he crossed them in numerous combinations and carefully recorded the outcomes. In doing so, he discovered that not all genes were equally potent. When he crossed a tall pea with a short one, the hybrid was always tall. Thus, he determined that the 'dominant' tall gene always held sway over the 'recessive' short gene.

However, when it came to breeding from the hybrids, the recessive gene came into its own. The result of crossing two tall hybrids was that three-quarters of them would produce tall offspring: these were the ones in which the gene combinations were tall × tall, tall × short and short × tall. The remaining quarter, where two short genes had combined, would produce short peas. Comparable results occurred with other characteristics, such as colour, leaf and pod formation.

Mendel had effectively invented the science of genetics, even if the term did not come into use until 1906. Although he published his results in 1866, their significance was not recognised in his lifetime. In 1868 he was elected abbot of the monastery and had less time for science. He died in 1884, and it was another 16 years before three European botanists conducted similar experiments and, on searching the literature, recognised that Mendel had covered the ground before them.

By coincidence, it was at about this time that Fairchild's contribution to horticultural history was first formally recognised. By now, hybridising had come to be accepted by most gardeners as the preferred way of improving plant quality. As an anonymous writer in the *Gardener's Chronicle* in January 1881 observed, in words that presaged the controversy over genetic modification more than a hundred years later: 'Hybridising was formerly regarded as a sacrilegious subversion of nature, and those who practised the art were stigmatised as mischievous intermeddlers in the works of the Creator.' Some writers and botanists 'would have consigned all hybrids to the rubbish heap as being of impure descent'. All the same, 'gardeners did not stay their hands in the work of rearing novelties, heedless of the "confusion" they were causing'.

On 11 July 1899 the Royal Horticultural Society organised its first conference on hybridisation and crossbreeding, held at its gardens at Chiswick in West London. It was attended both by botanists and by practical gardeners and hybridisers, and an undertone to the speeches suggested that there was tension between the two groups, the gardeners feeling that their success in breeding hybrids had not been matched by any advance in understanding by the scientists of the mechanisms they were employing.

The conference was opened by its chairman, Dr Maxwell Masters, a Fellow of the Royal Society, whose speech was reported at length in *The Times* on the following day. 'We are met,' he began, 'to discuss one of the most important problems of modern progress in experimental horticulture.' In practical terms, he noted, hybridising techniques had not greatly advanced for generations, even though nine-tenths of modern plants 'were the productions of the gardener's art, and not natural productions'.

He paid tribute to Fairchild as the first person to form a hybrid artificially: 'You will remark that this was more than

40 years before Kölreuter began his elaborate series of experiments.' Praising the bequest for the Shoreditch Lecture, Dr Masters observed that in Fairchild's time there had been a strong prejudice against hybridisation among religious people:

> It was said that by crossbreeding plants, people were flying in the face of providence and that the process was wicked . . . an impious interference with the laws of nature. But Dean Herbert [of Manchester, 1779–1847] showed that by crossing two species of daffodils which he found in the Pyrenees, he could produce flowers similar to those which abounded in the locality; and he therefore argued that if nature did the same thing he must not be blamed for doing what nature did.

Dr Masters added that the prejudice against hybridisation had been carried so far that nurserymen were afraid to exhibit hybrids at RHS shows 'because they might injure the feelings of some over-sensitive religious persons; and they therefore exhibited them as wild species from abroad'. More surprisingly, there was now a prejudice against hybrids among some botanists:

> It is not indeed altogether surprising that the botanists should have objected to the inconvenience and confusion introduced into their pretty little systems of classification by the introduction of hybrids and mongrels, and that they should object to hybrid species, and much more to hybrid genera. But it would be very unscientific to prefer the interests of our systems to the extension of truth.

He was confident that the delegates to the conference 'will be not only advancing science, but also adding enormously to the welfare of humanity'.

Almost exactly 100 years later, on 15 July 1999, a rather different conference was held in the lecture room of the Jodrell Laboratory at Kew Gardens, not at all far from Chiswick. Some 70 people, many of them writers who specialised in gardening, had met to discuss an issue that had been gaining more and more coverage in the press and the broadcast media during the previous nine months – but coverage that had generated a lot more heat than light.

In the century that separated the two meetings, scientists had made enormous strides in understanding the complex structure and functions of genes in determining the characteristics of living things. It was discovered that genes were made of a chemical called DNA (deoxyribose nucleic acid), in which genetic information was stored as a code, to be passed on to future generations. With this greater understanding had come, towards the end of the century, the ability to interfere in these functions, to manipulate them in a far more sophisticated way than the simple transference of pollen to a flower's pistil. DNA can be reproduced and replicated, and genes known to provide protection against serious disease, or to perform other desirable functions, can be inserted into living organisms. Among the products of this new biotechnology have been effective drugs and vaccines, as well as the much-publicised cloning of a sheep called Dolly.

In agriculture and horticulture, genetic engineering has mainly been used to breed crops with high productivity and disease and pest resistance, in effect building on the advances that have been made in these areas by hybridisation throughout the twentieth century. There have also been experiments in breeding fruit and vegetables with better colour and flavour, higher starch content (to increase nutritional value) and a longer shelf life, as well as producing flowers in colours that hybridisation has been unable to achieve.

Many scientists believe that we are at the limit of what can be done by the non-genetic methods that Fairchild pioneered. The crucial difference between the two techniques is that while hybridisation can occur only between species that are closely related, genes from almost any living entity can theoretically be patched into any other – even animal genes into vegetables, and vice versa. It is this possibility that has led to lurid scares about the creation of Frankenstein-style monsters with the capacity to wreak havoc on the environment and on the existing order. The Prince of Wales has been the highest-profile opponent of experiments with genetic modification, maintaining that we 'meddle with nature at our peril'.

By 1999 several genetically engineered crops were being produced, mainly in the United States, with soybeans and oilseed rape being the most frequent subjects of the experiments. In most cases, the purpose of the genetic modification was to breed into the crops a resistance to certain weedkillers. This would mean that they could be sprayed to kill the weeds around them and thus increase yields. Critics of the procedure pointed out that in many cases the manufacturer of the weedkiller was the same company that was doing the research on the genetically modified crops, and that the principal motive could therefore be commercial profit rather than scientific advance.

They had two other worries, more directly involving the environment. One was that the artificially inserted genes could be passed on through cross-fertilisation to the weeds themselves, making them immune to existing weedkillers and becoming super-weeds, matching the ability of such plants as Japanese knotweed to colonise large acreages very quickly. This would set the stage for the introduction of even more powerful substances to deal with them – a toxic snowball effect. The other fear was that by spraying weedkiller on to the immune crops you would kill not only the targeted weeds but related native plants, thus reducing what has come to be called biodiversity.

To answer the concern about cross-fertilisation, scientists developed a 'terminator gene' that would prevent modified crops from producing viable seed. This, too, was criticised because it would mean that farmers would have to buy fresh seed every year – a distinct drawback for those in the developing countries who were supposed to be among the main beneficiaries of the new techniques. Another form of genetic modification, the insertion of resistance to infestation by pests, was promoted as meaning that the use of chemical pesticides could be reduced, but this did not satisfy the environmentalists either: they said it would unbalance the environment by cutting the numbers of insects that provided food for birds and mammals.

The wave of concern that gave rise to the Kew conference had begun in August 1998, when Dr Arpad Pusztai, of the Rowett Research Institute in Aberdeen, said on television that, as part of a research project, he had fed genetically modified potatoes to rats for ten days. He maintained that, as a result, their immune systems had been weakened and some of their organs had shrunk – including the kidney, the spleen and the brain. His findings were subsequently challenged and he left the Institute; but other scientists supported him, and the issues he had raised refused to go away. Opinion polls showed that a majority of people had serious doubts about GM foods, and in the summer of 1999 groups of activists, calling for the suspension of all trials of the procedure, made several early-morning raids on fields in which GM crops were being tested, causing serious damage and hindering the experiments.

The July conference had been arranged by the Garden Writers Guild, whose members were finding that the issue of GM crops and plants was increasingly impinging on their work, in the form of worried questions from readers and editors. The Prince of Wales was not there, but he was represented by David Howard, head gardener at his organically run estate at Highgrove in Gloucestershire.

The first speaker was Professor Geoffrey Dixon from the University of Strathclyde, managing director of a consultancy on plant science called GreenGene International. In advance of the meeting he had circulated a paper called 'Genetically modified organisms – the science behind the headlines'. While advocating 'rigorous, transparent, independent testing of the implications for human health and welfare and for environmental impact', the paper summed up the principal argument for GM experiments in a paragraph: 'Plant breeding using non-genetic modification has increased the yields of staple crops in the past 50 years by factors up to ten-fold but has now reached a plateau. Yet the world's population continues to grow, and the amount of cultivable land is diminishing.' He added that an advantage of genetic modification for plant breeders was that it 'makes breeding a prescriptive rather than a descriptive discipline'.

And that, to nobody's real surprise, was what worried

many at the Kew meeting. It did not take long for them to begin to express their concerns. In the opening panel discussion, there was breezy talk from plant breeders about what genetic modification could mean for flower growers: more and cheaper F1 hybrids, new colour ranges for popular plants – blue roses and mauve carnations were promised – as well as flowers that would last longer in vases and slow-growing grasses for local authorities that could not afford frequent mowing.

Then came the question-and-answer session, at which 'meddling with nature' was quickly and repeatedly raised by a vociferous element in the audience who passionately opposed GM experiments. They said they were prepared to forgo all the promised benefits, so long as the price was the manipulation of systems that had been created by a higher authority. One of the experts pointed out that meddling with nature was what gardening was essentially about: the purpose of hybridisation, and of introducing plants from abroad, was to alter the range of things that could be grown in Britain naturally. But a panellist from the anti-GM pressure group Genewatch, responding to the question of why the organic movement was against genetic modification, said it was because the process worked against nature rather than with it. This was an argument that Fairchild and his contemporaries would have understood.

The summing-up was by Richard Dawkins, Professor of the Public Understanding of Science at Oxford University and the author of several popular books about genetics and evolution. He pointed out that genetic modification had been occurring slowly for millions of years – that is what evolution means. But it was a gradual and random process, with no sudden leaps, and that was what distinguished it from man-made GM techniques. As for meddling with

God's design, it had taken until the mid-nineteenth century to recognise conclusively that there was no such thing, that natural selection was not based on any creator's foresight or planning.

While he shared some of the worries about genetic modification, he deplored the hysterical tone of voice employed by some of its opponents and the 'emotional claptrap' they used to support their arguments. He did not like phrases such as 'playing God' and 'interfering with nature'. People who used them implied that nature had a kind of wisdom that humans lacked – yet nature had no foresight and was not universally benign; hence the incidence of plagues, famines and diseases. In principle, it was possible for humans to design things better, although it may be that the political institutions we had devised would not produce the results we wanted. His conclusion was that although, in the short term, progress could be impeded by the 'propaganda campaign' against GM, in the long run advanced technology could lead to some improvements in agriculture and horticulture. It was as important, though, to be aware of possible dangers as it was to resist being sentimental about nature.

Amid all the prejudice and emotion that the day's discussion unleashed, one challenge from the audience stood out for its simplicity: 'Who needs a blue rose?' The question 'Who needs Fairchilds' mule?' would have been equally valid had the meeting been held in the 1720s. Transferring genes between widely disparate plant material, and even from animals to vegetables, is certainly a long step beyond what has been done before: but then Fairchild's insertion of the pollen of the sweet william into a carnation was a long step beyond the former practice of letting naturally compatible species breed spontaneously.

So who needs it? Nobody who has to ask the question can conceivably identify with Thomas Fairchild and the other gardening pioneers, botanists and plant-hunters, whose quest for horticultural novelty drove their life's work – even overriding, in Fairchild's case, a deeply held religious faith. For them, a blue rose would have been something, if not to die for, at least to pluck, wrap carefully and take along to Crane Court for the curious members of the Royal Society to marvel at.

Select Bibliography

Books
(published in London except where stated)

Written before 1900
Blair, Patrick, *Botanick Essays*, 1720
Bradley, Richard, *The Monthly Register of Experiments and Observations in Husbandry and Gardening, with A General Treatise of Husbandry and Gardening*, 1724
Bradley, Richard, *New Improvements of Planting and Gardening, both Philosophical and Practical*, 1726
Campbell, R., *The London Tradesman*, 1747
Cowell, John, *The Curious and Profitable Gardener*, 1730
Defoe, Daniel, *A Tour through the Whole Island of Great Britain*, 1726; edited and with notes by Pat Roger, Penguin, 1971
Ellis, Henry, *History and Antiquities of the Parish of St Leonard, Shoreditch*, 1798
Evelyn, John, *Diary*, G. Newnes, 1903 edition
——*The Writings of John Evelyn*, edited by Guy de la Bédoyère, Boydell Press, 1995
Fairchild, Thomas, *The City Gardener*, 1722
Grew, Nehemiah, *The Anatomy of Plants*, 1682
Hazlitt, W. C., *Gleanings in Old Garden Literature*, Elliot Stock, 1887
Humphreys (translator), *Spectacle de la Nature, or Nature Displayed*, 1740
Johnson, George W., *A History of English Gardening*, 1829

London, George, and Wise, Henry, *The Retir'd Gardener* (two vols), 1706

Meager, Leonard, *The New Art of Gardening*, 1697

Miller, Philip, *The Gardener's and Florist's Dictionary*, 1724

——*The Gardener's Dictionary*, 1731

Nichols, John, *Illustrations of the Literary History of the Eighteenth Century* (eight vols), 1817–58

Pulteney, Richard, *Sketches of the Progress of Botany in England from Its Origin to the Introduction of the Linnaean System* (two vols), 1790

Society of Gardeners, *Catalogus Plantarum*, 1730

Stow, John, *The Survey of London 1598–1603*, Dent: Everyman, 1987

Switzer, Stephen, *Ichnographia Rustica, or The Nobleman, Gentleman and Gardener's Recreation* (three vols), 1715

——*The Practical Fruit Gardener*, 1724

Weld, C. R., *History of the Royal Society* (two vols), 1848

Whitmill, Benjamin, *Kalendarium Universale, or the Gardener's Universal Calendar*, 1726

Written after 1900

Amherst, Alicia, *London Parks and Gardens*, Constable, 1907

Barnes, Melvyn, *Root and Branch, a History of the Worshipful Company of Gardeners*, Gardeners' Company, 1994

Besant, Sir Walter, *London in the Eighteenth Century*, A. & C. Black, 1902

Bisgrove, Richard, *The National Trust Book of the English Garden*, Viking, 1990

Brazell, J. H., *London Weather*, Meteorological Office, 1968

Bristow, Alec, *The Sex Life of Plants: A Study of the Secrets of Reproduction*, Barrie & Jenkins, 1979

Brock, C. Helen, *Dr James Douglas's Papers and Drawings in the Hunterian Collection, Glasgow University Library*, Wellcome

Unit for the History of Medicine, University of Glasgow, 1994
Brooks, E. St John, *Sir Hans Sloane: The Great Collector and His Circle*, Batchworth, 1954
Clokie, Hermia Newman, *An Account of the Herbaria of the Department of Botany in the University of Oxford*, Oxford University Press, 1964
Coats, Alice M., *Flowers and Their Histories*, Hulton, 1956
——*Garden Shrubs and Their Histories*, Vista, 1963
Crane, Maurice A., *The Aldbourne Chronicle*, published by author, Aldbourne, 1974
Crossweller, W. T., *The Gardeners' Company, 1605–1907*, Gardeners' Company, 1908
Dandy, J. E., *The Sloane Herbarium*, British Museum, 1958
Desmond, Ray, *Dictionary of British and Irish Botanists and Horticulturalists*, Taylor & Francis, 1994
Earle, Peter, *The Making of the English Middle Class*, Methuen, 1989
——*A City Full of People*, Methuen, 1994
Fisher, John, *The Origin of Plants*, Constable, 1982
Frick, George Frederick, and Stearns, Raymond Phineas, *Mark Catesby, the Colonial Audubon*, University of Illinois Press, Urbana, 1961
Gandy, Ida, *The Heart of a Village*, Moonraker Press, Aldbourne, 1975
Green, David, *Gardener to Queen Anne: Henry Wise (1653–1738) and the Formal Garden*, Oxford University Press, 1956
Hadfield, Miles, *A History of British Gardening*, Hutchinson, 1960
——Harling, Robert, and Highton, Leonie, *British Gardeners: A Biographical Dictionary*, Zwemmer, 1980
Harvey, John H., *Early Nurserymen*, Phillimore, 1974
Henrey, Blanche, *British Botanical and Horticultural Literature*

before 1800 (three vols), Oxford University Press, 1975
——*No Ordinary Gardener: Thomas Knowlton*, British Museum (Natural History), 1986

Hessayon, David, *The Armchair Book of the Garden*, Century Publishing, 1983

Holmes, Geoffrey, *Politics, Religion and Society in England, 1679–1742*, Hambledon Press, 1986

——(ed.), *Britain after the Glorious Revolution, 1689–1714*, Macmillan/St Martin's Press, 1969

Hoyles, Martin, *The Story of Gardening*, Journeyman Press, 1991

——*Gardeners' Delight: Gardening Books from 1560 to 1960, vol. 1*, Pluto Press, 1994

——,*Bread and Roses: Gardening Books from 1560 to 1960, vol. 2*, Pluto Press, 1995

Jones, M. G., *The Charity School Movement: A Study of 18th-Century Puritanism in Action*, Cambridge University Press, 1938

LCC Survey of London, vol. VIII, Parish of St Leonard, Shoreditch, LCC/Batsford, 1922

Le Rougetel, Hazel, *The Chelsea Gardener: Philip Miller, 1691–1771*, Natural History Museum, 1990

Moreton, C. Oscar, *Old Carnations and Pinks*, Rainbird/Collins, 1955

Pasti, George Jr., *Consul Sherard: Amateur Botanist and Patron of Learning, 1659–1728*, unpublished thesis, University of Illinois, 1950

Picard, Liza, *Restoration London*, Weidenfeld, 1997

Roberts, H. F., *Plant Hybridization before Mendel*, Princeton University Press, 1929

Steele, Arnold F., *Worshipful Company of Gardeners of London: A History of Its Revival, 1890–1960*, Gardeners' Company, 1964

Stevenson, John (ed.), *London in the Age of Reform*, Blackwell, 1977

Thomas, Keith, *Man and the Natural World: Changing Attitudes in England 1500–1800*, Allen Lane, 1983

Walters, S. M., *The Shaping of Cambridge Botany*, Cambridge University Press, 1981

Weinreb, Ben, and Hibbert, Christopher (eds), *The London Encyclopaedia*, Macmillan, 1983

White, Michael, *Isaac Newton: The Last Sorcerer*, Fourth Estate, 1997

Zirkle, Conway, *The Beginnings of Plant Hybridization*, University of Pennsylvania, 1935

Articles and Pamphlets

Brock, C. Helen, 'James Douglas (1675–1742), Botanist', *Journal of the Society of Bibliography of Natural History*, vol. 9, no. 2, 1979

Cowie, L. W., 'Bridewell', *History Today*, vol. XXIII, 1973

Egerton, Frank N. 3rd, 'Richard Bradley's Illicit Excursion into Medical Practice in 1714', *Medical History*, vol. XIV, 1970

——'Richard Bradley's Relationship with Sir Hans Sloane', *Notes and Records of the Royal Society*, 25 June 1970

Fairchild, Thomas, 'New Experiments Relating to the Different and Sometimes Contrary Motion of the Sap in Plants', *Royal Society Philosophical Transactions 1719–1733*, vol. 33, no. 384

Faircloth, Nick, 'Pursuing a Passionate Desire' (re Mark Catesby), *The Garden*, vol. 123, part 11, Nov. 1998

Gorer, Richard, 'Nurseries, Hybrids and Plant Collectors', *Archives* (Journal of the British Records Association), vol. XII, no. 55, spring 1976

Griffiths, Mark, 'Vessels of the Crown', *The Garden*, vol. 124, part 5, May 1999

Harvey, John H., 'The Stocks Held by Early Nurseries', *The Agricultural History Review*, vol. 22, 1974

——'Leonard Gurle's Nurseries and Some Others', *Garden History*, vol. 3, no. 3, 1975

——'Mid-Georgian Nurseries of the London Region', *Transactions of the London and Middlesex Archaeological Society*, vol. 26, 1975

——'The Nursery Garden', Museum of London, 1990

Le Lièvre, Audrey, 'Hoxton's Horticulturalist', *Country Life*, vol. 182, no. 44, 3 Nov. 1988

Roberts, W., 'R. Bradley, Pioneer Garden Journalist', *Journal of the Royal Horticultural Society*, vol. LXIV, 1939

——'John Renton and Other Hoxton Nurserymen', *Journal of the Royal Horticultural Society*, vol. LXIII, 1939

Thomas, H. Hamshaw, 'The Rise of Natural Science in Cambridge', *Cambridge Review*, 28 May 1937

——'Richard Bradley, an Early 18th-Century Biologist', *Bulletin of the British Society for the History of Science*, vol. 1, no. 7, May 1952

Willson, E. J., 'The Records of Nurserymen', *Archives*, vol. XII, no. 55, spring 1976

Index

(Numbers in **bold** type refer to illustrations. Those on the unnumbered colour pages are indicated by **pl.**)

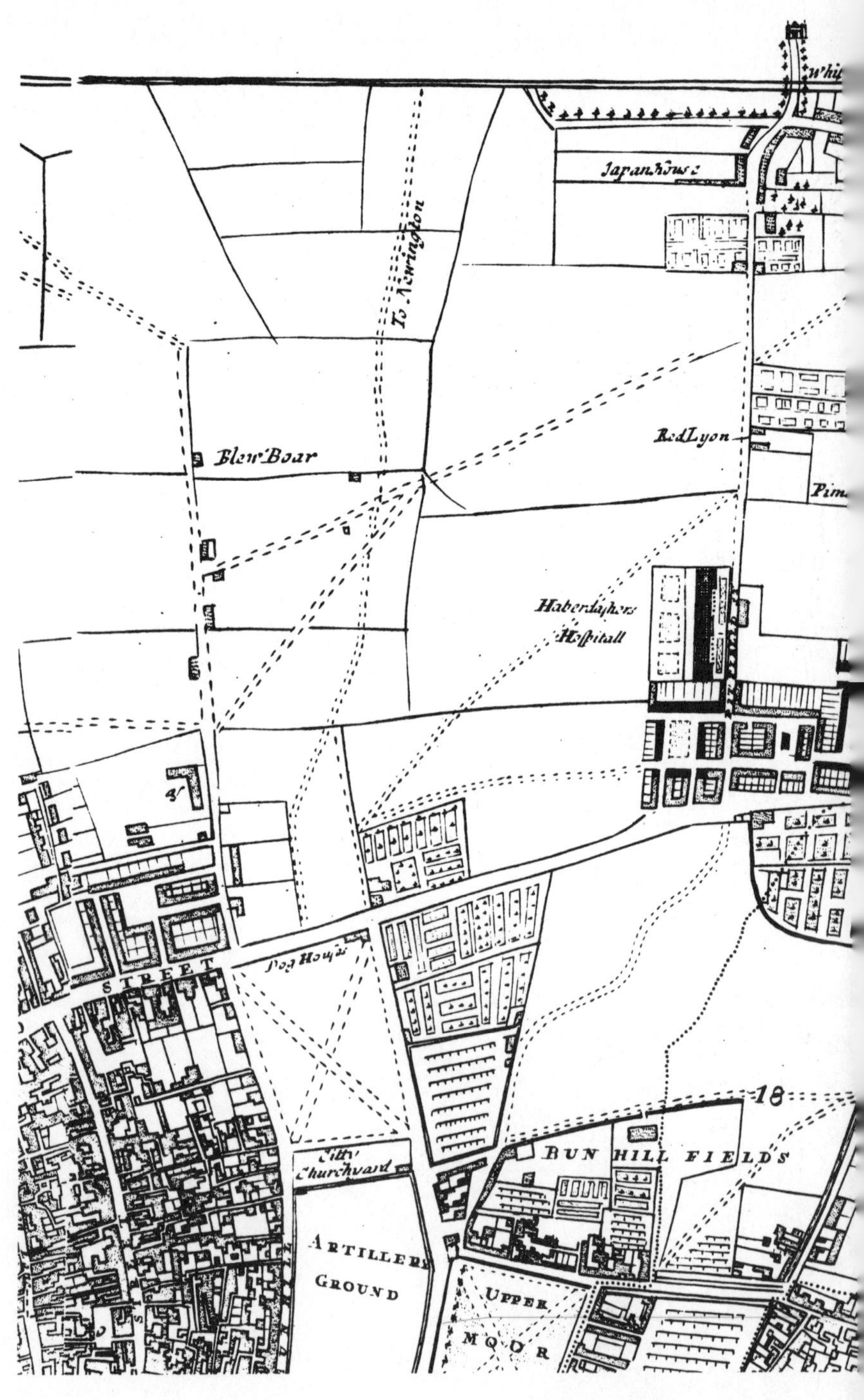
Japan house
To Newington
Blew Boar
Red Lyon
Haberdashers Hospitall
Dog House
STREET
18
Citty Churchyard
BUN HILL FIELDS
ARTILLERY GROUND
UPPER MOOR
SOUT